One Ewe over the Cuckoo's Nest

One Ewe over the Cuckoo's Nest

Kathryn Lamb

FARMING PRESS

FARMING PRESS

First published 1991

ISBN 0 85236 227 7

A catalogue record for this book is available
from the British Library

Published by Farming Press Books
4 Friars Courtyard, 30-32 Princes Street
Ipswich IP1 1RJ, United Kingdom

Distributed in North America
by Diamond Farm Enterprises,
Box 537, Alexandria Bay, NY 13607, USA

Cover design by Andrew Thistlethwaite
Phototypeset by Typestylers, Ipswich
Printed and bound in Great Britain by Redwood Press, Melksham, Wilts.

Sheep Laughs

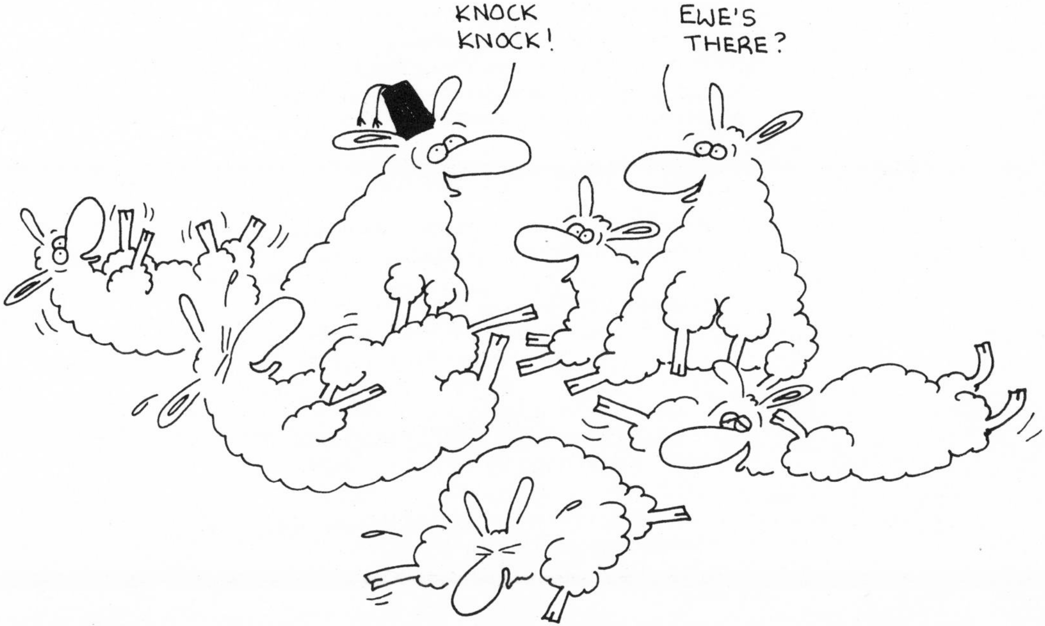
KNOCK KNOCK!
EWE'S THERE?

BATTERING RAM

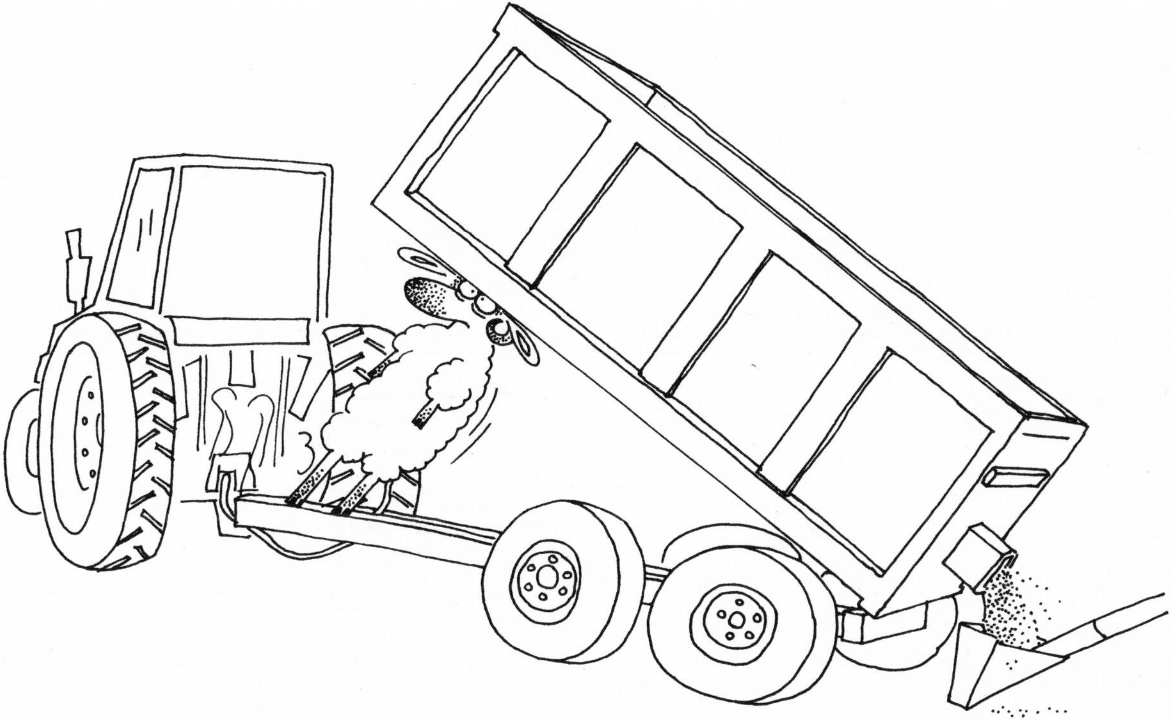

HYDRAULIC RAM

LITTLE BO PEEP HAS LOST HER SHEEP,
AND I KNOW WHERE TO FIND THEM,
THEY'RE IN THE DEEP FREEZE
  WITH PACKETS OF PEAS
    ALL NEATLY STACKED BEHIND THEM!

SHEEP   WORRYING

4

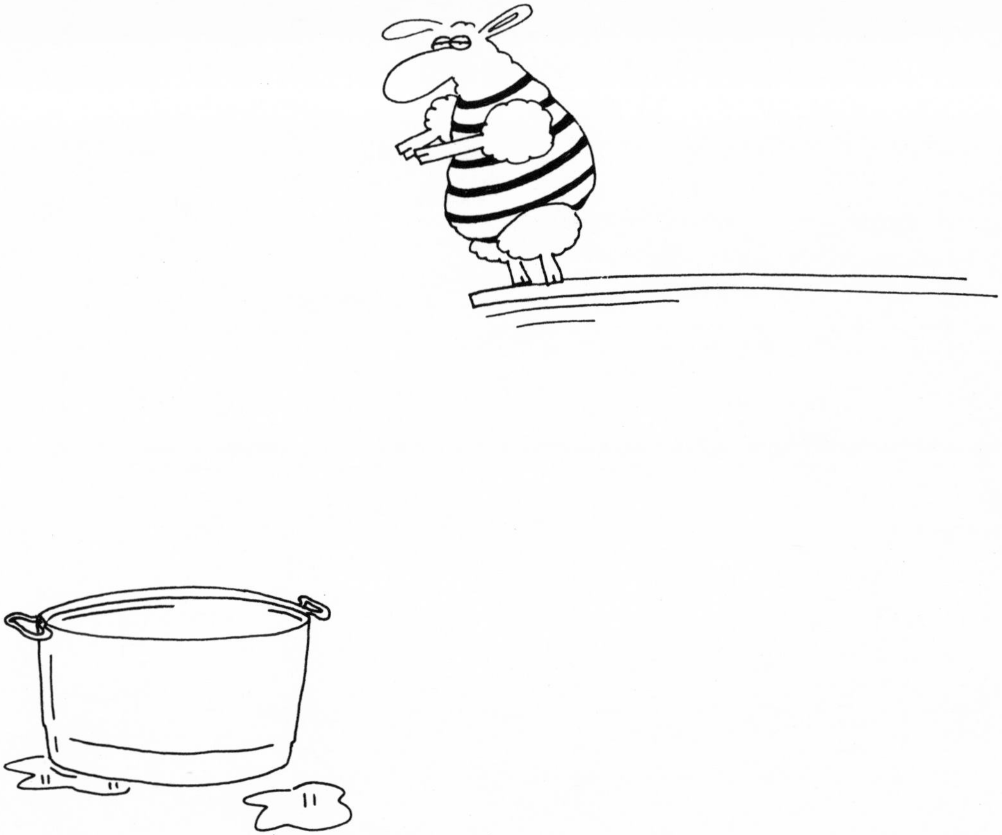

SHEEP DIP

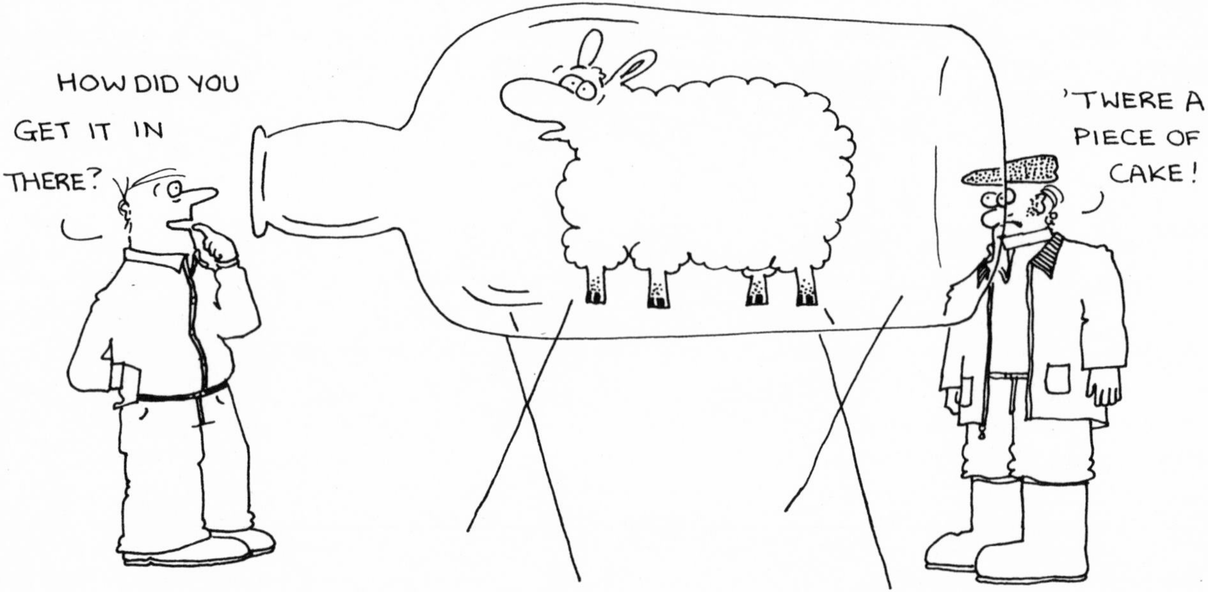
HOW DID YOU GET IT IN THERE?
'TWERE A PIECE OF CAKE!

SHEEP IN A BOTTLE

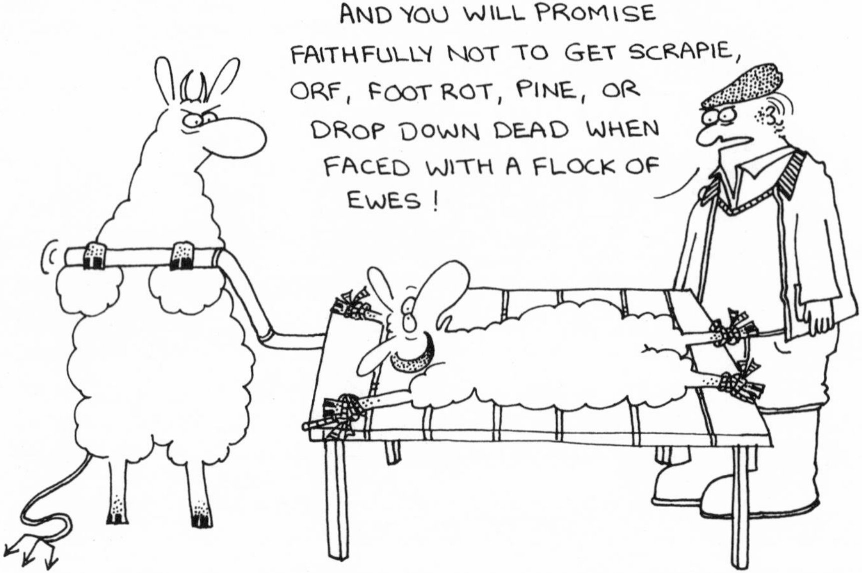
AND YOU WILL PROMISE FAITHFULLY NOT TO GET SCRAPIE, ORF, FOOT ROT, PINE, OR DROP DOWN DEAD WHEN FACED WITH A FLOCK OF EWES !

RACK OF LAMB

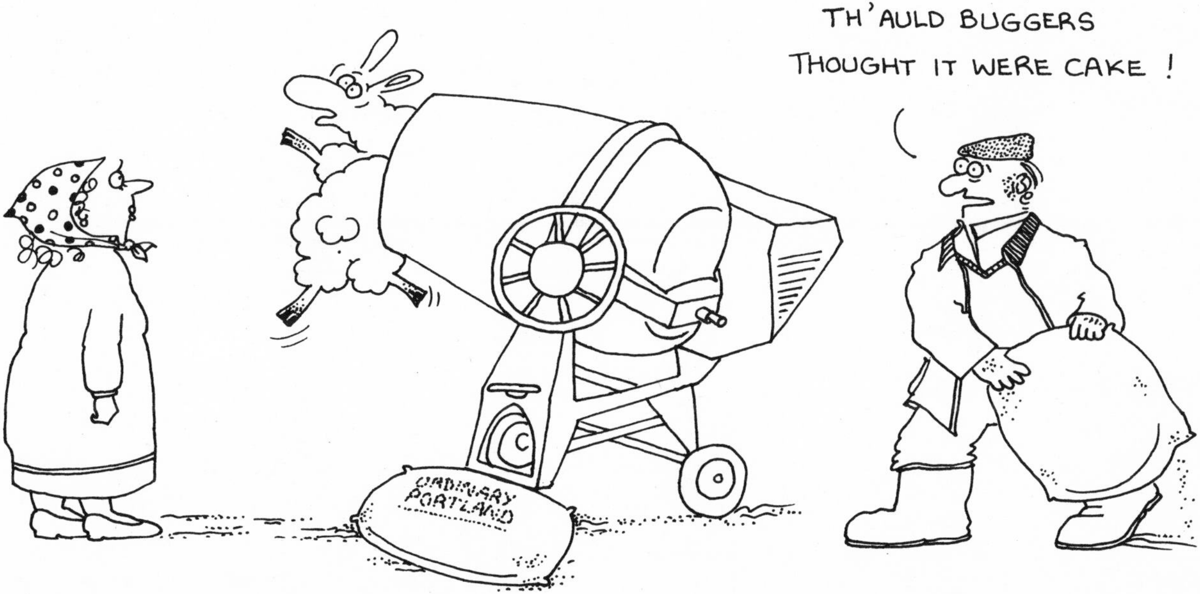
TH'AULD BUGGERS
THOUGHT IT WERE CAKE !
ORDINARY
PORTLAND

RAM PANT

ONE UP RAM SHIP

11

EWE F. O.

13

LAMB FATALE

RAMIFICATIONS

15

LIFE BEHIND BAAS

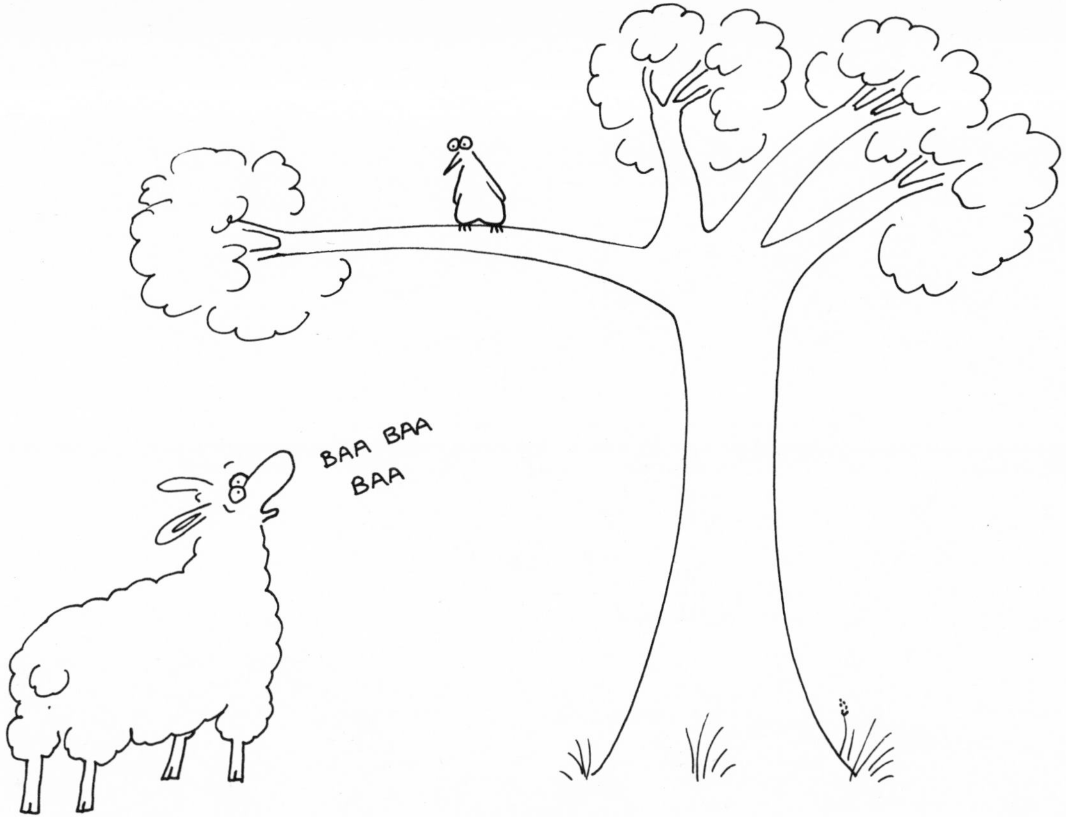
BAA BAA
BAA

BAAING UP THE WRONG TREE

SHEEP SORTER

19

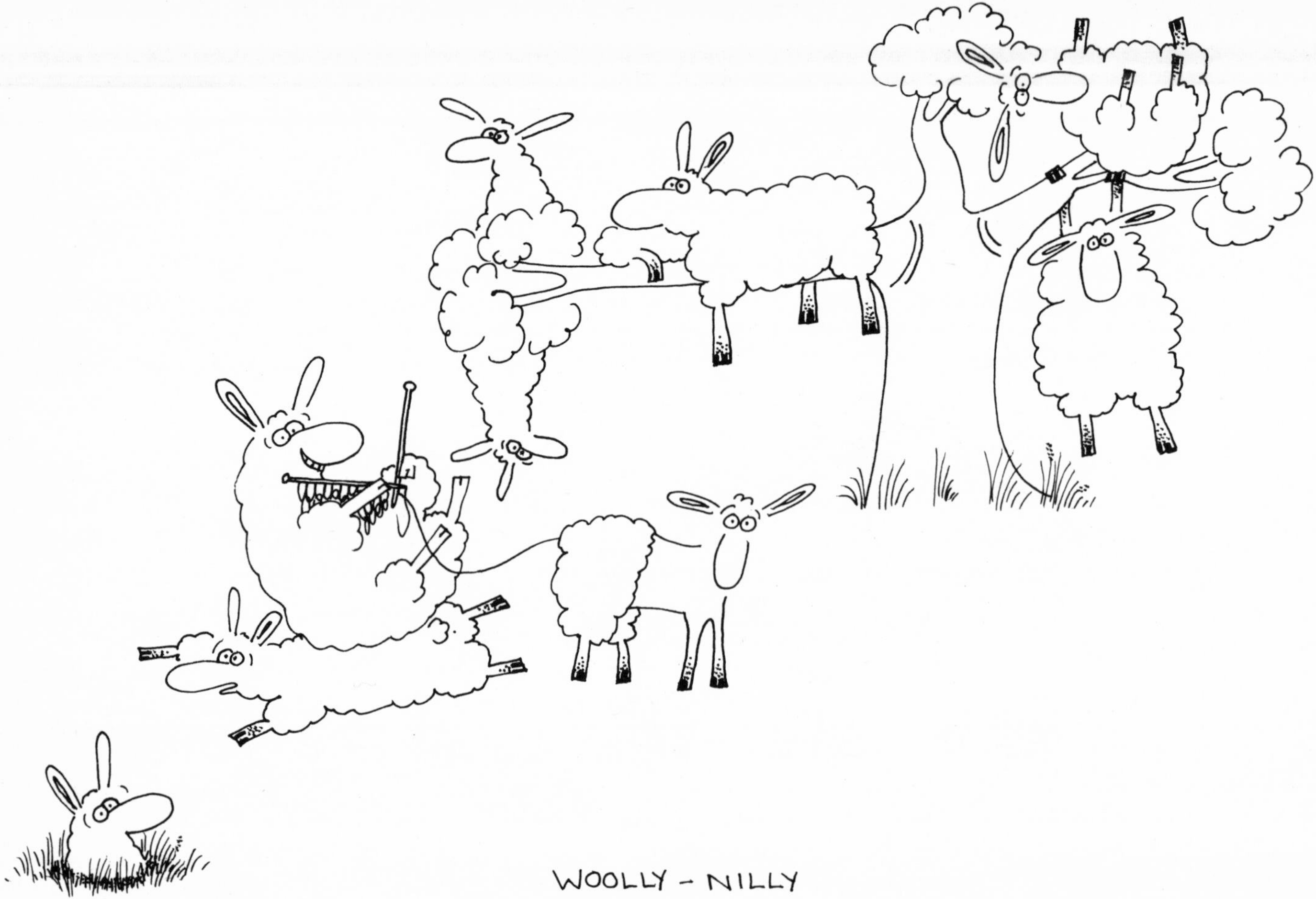

WOOLLY - NILLY

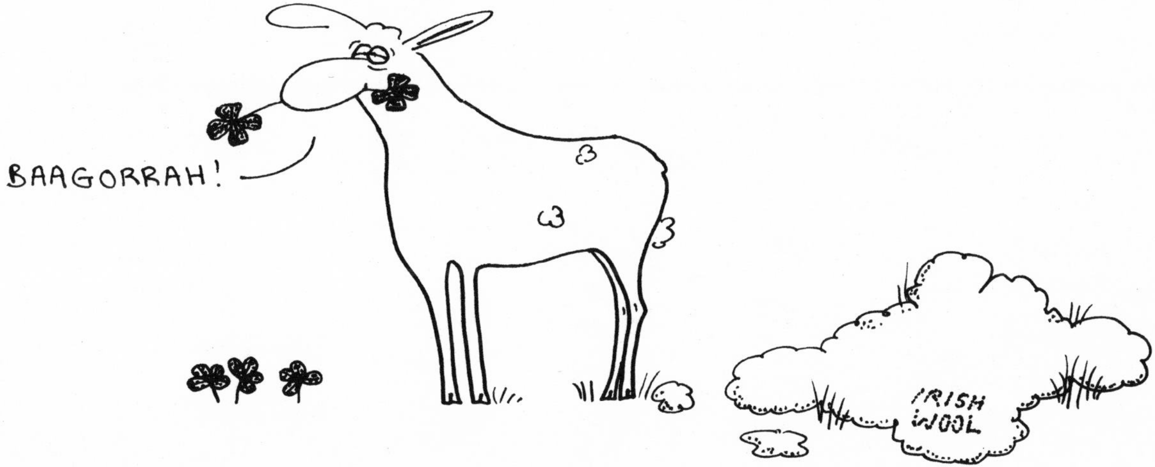
BAAGORRAH!
IRISH WOOL

SHORN O' SHEEP

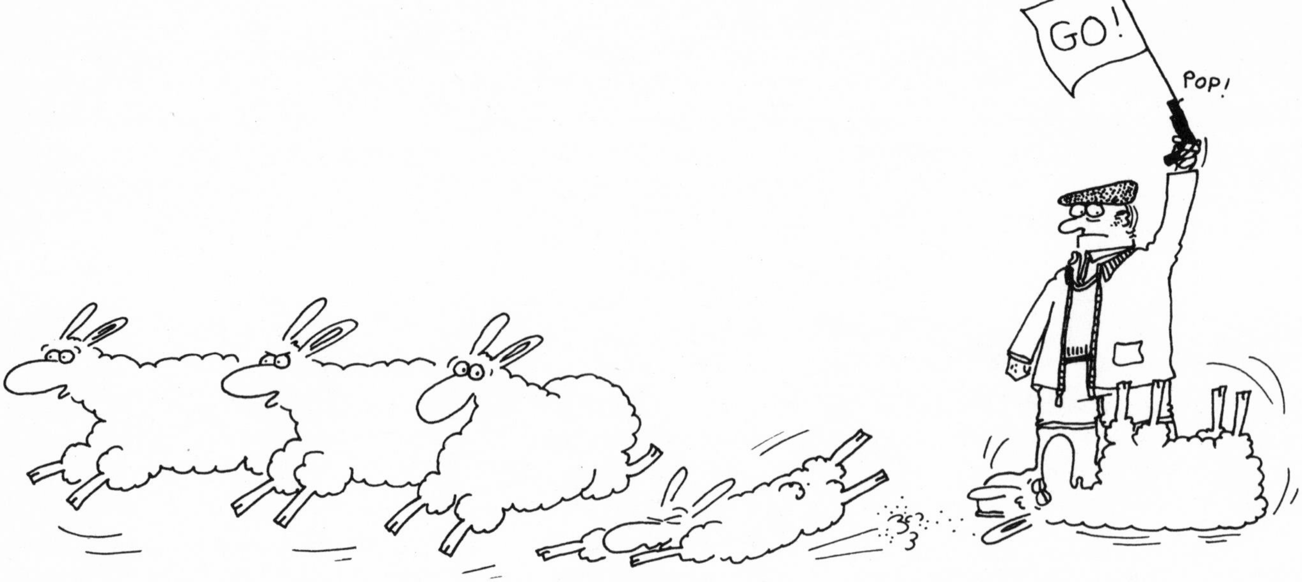
GO!
POP!

SHEEP RACE

TALL SHEEPS RACE

LAMBRETTA

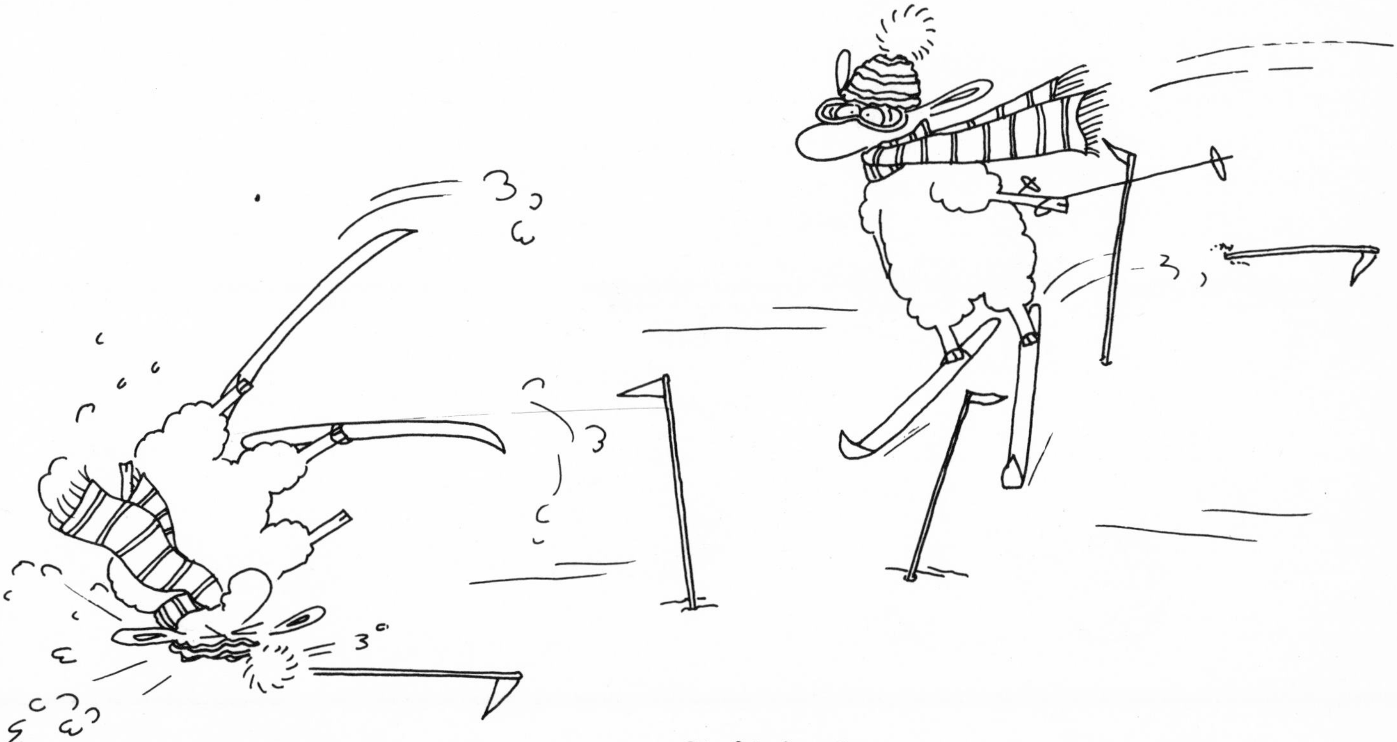

SLALAMB

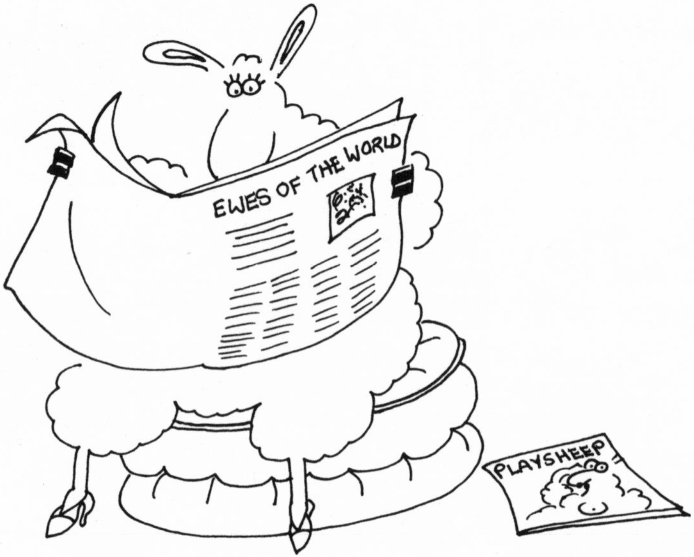
EWES OF THE WORLD
PLAYSHEEP

EWESPAPER

BAA BAA
BAA BAA BAA
BAA... ER... BAA
BAA BAA BAA
BAA . . .
EWES
AT
TEN

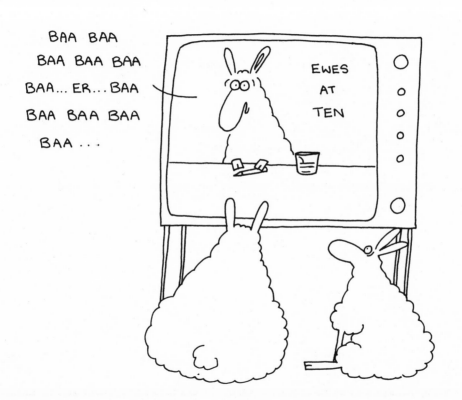

READING FROM THE AUTOCEWE

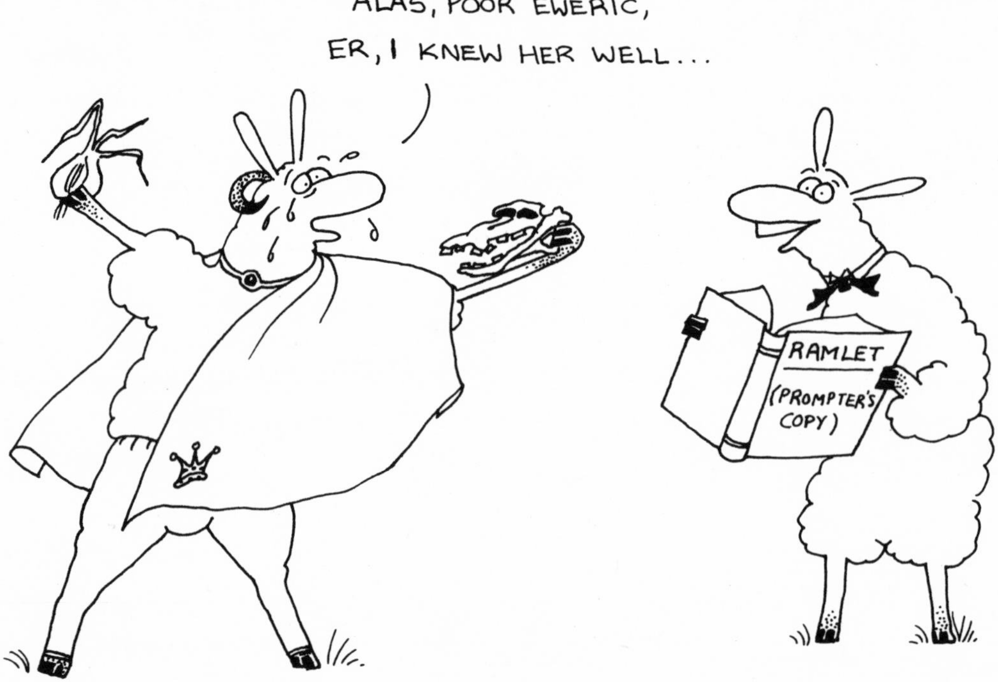
ALAS, POOR EWERIC,
ER, I KNEW HER WELL...
RAMLET
(PROMPTER'S COPY)

THE REALLY EWESLESS THEATRE COMPANY

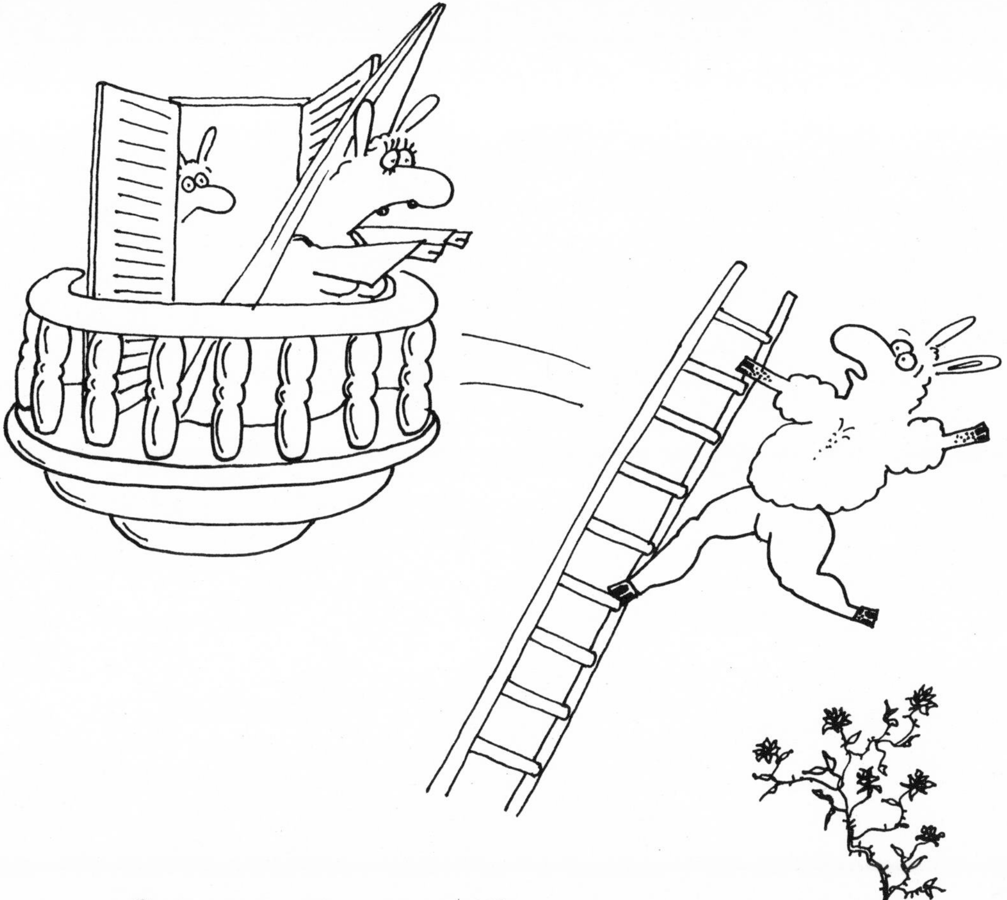

RAMEO AND JEWELIET

SOOTY AND SHEEP

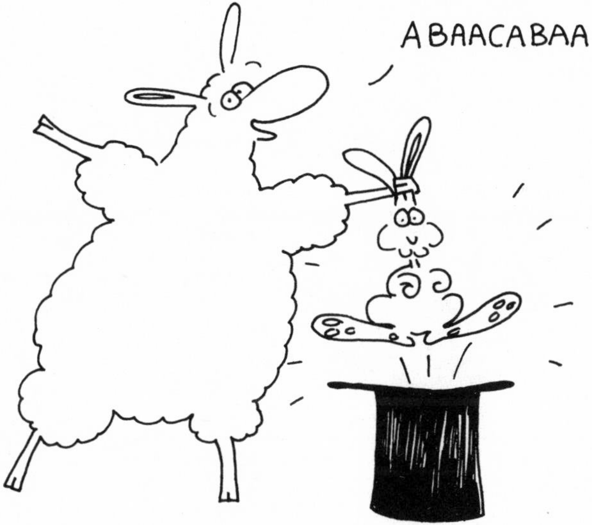
ABAACABAARA!

A SHEEP TRICK

GOLDILOCKS AND THE THREE BAAS

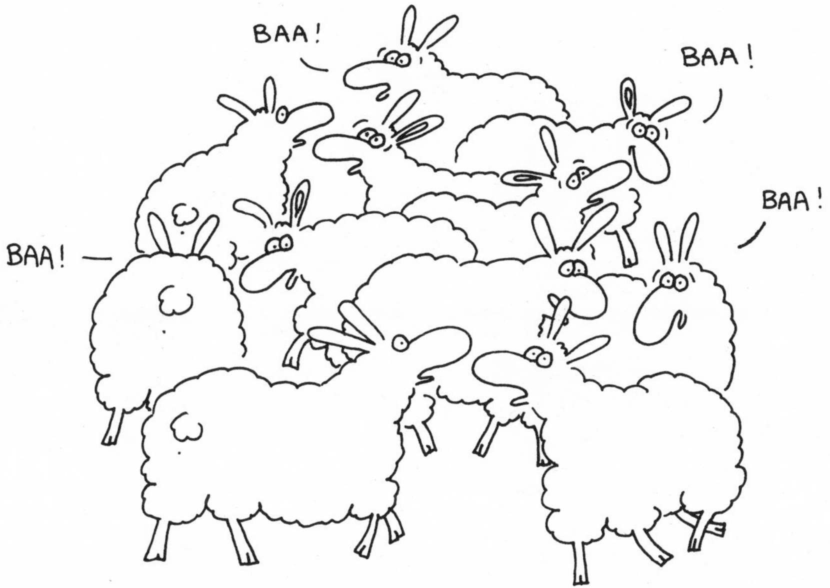
BAA!
BAA!
BAA!
BAA!

BAA FROM THE MADDING CROWD

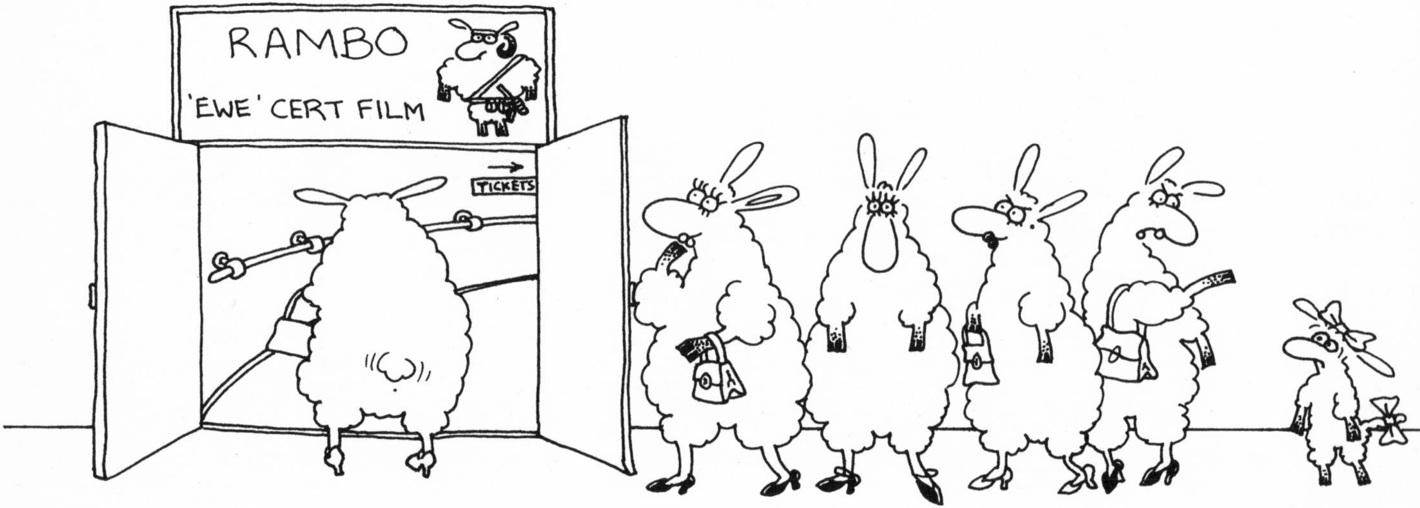
RAMBO
'EWE' CERT FILM
TICKETS

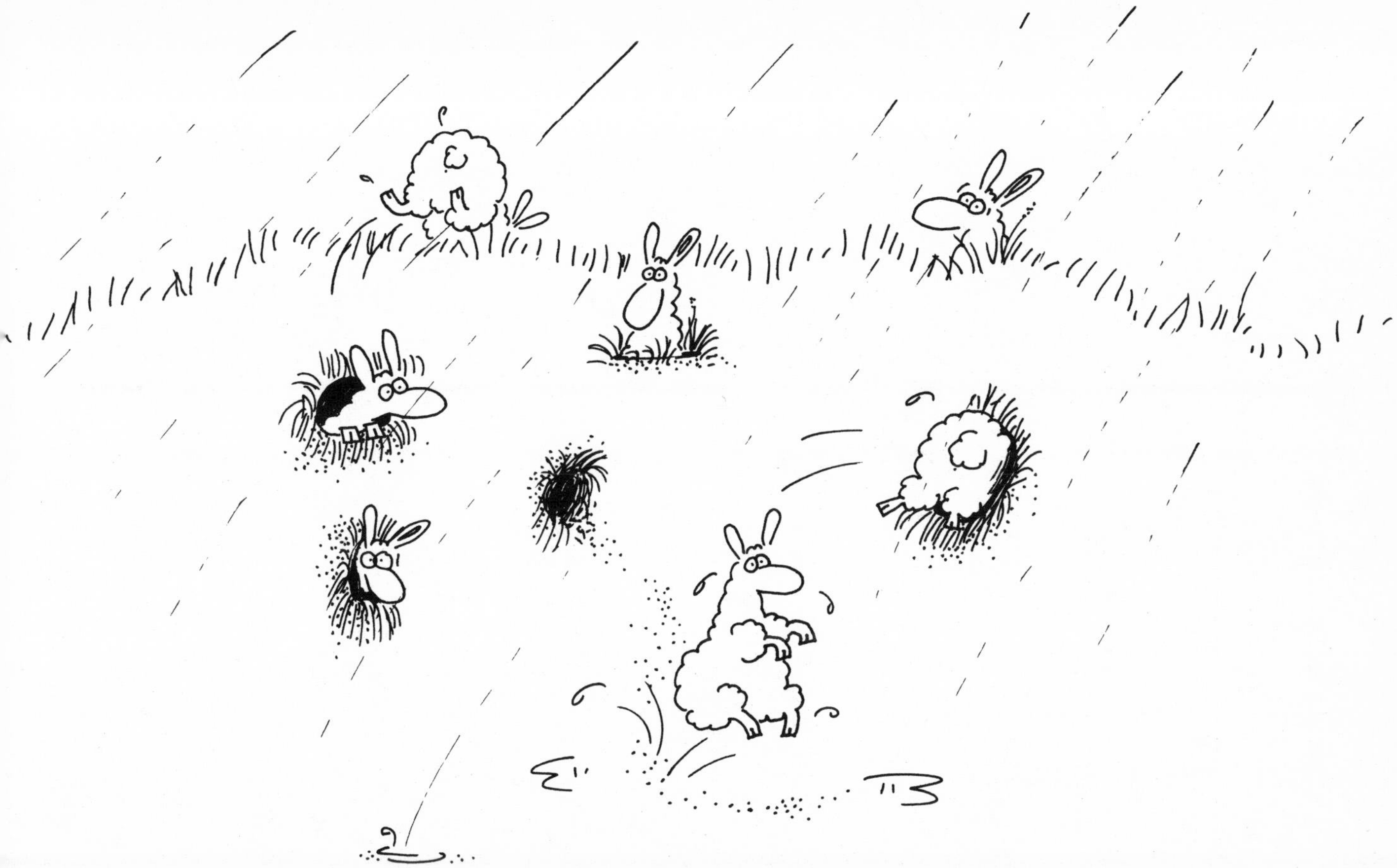

WATERSHEEP DOWN

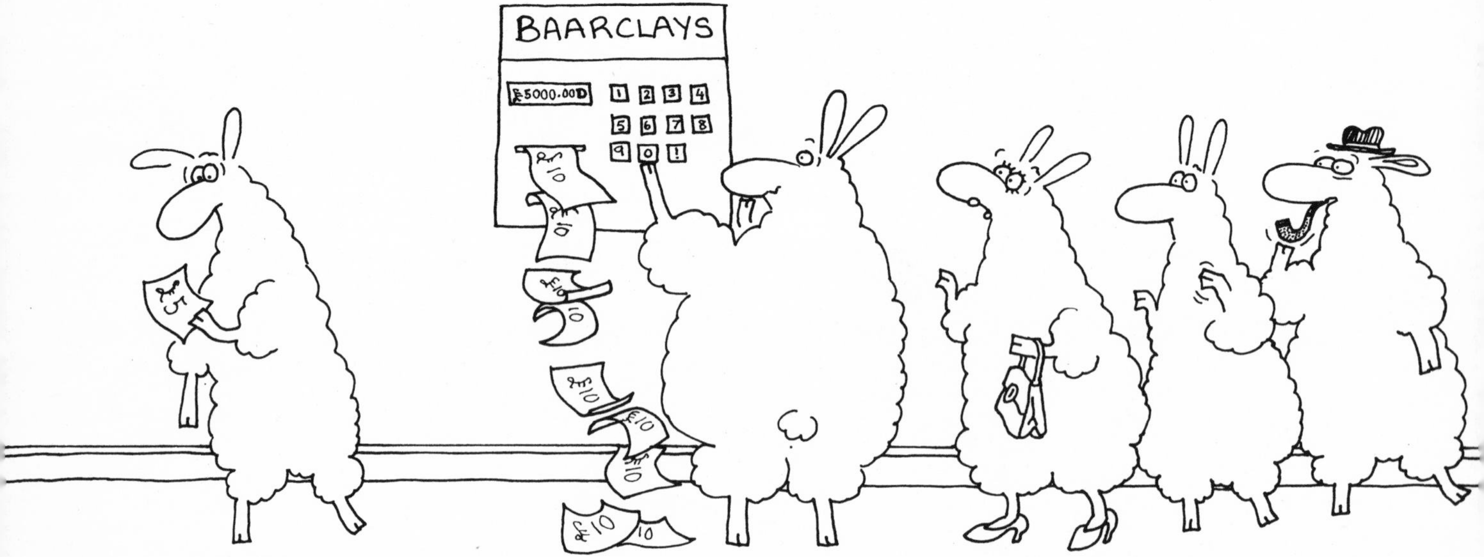
BAARCLAYS
£5000.00D
1 2 3 4
5 6 7 8
9 0 !
£10

LAMB BANK

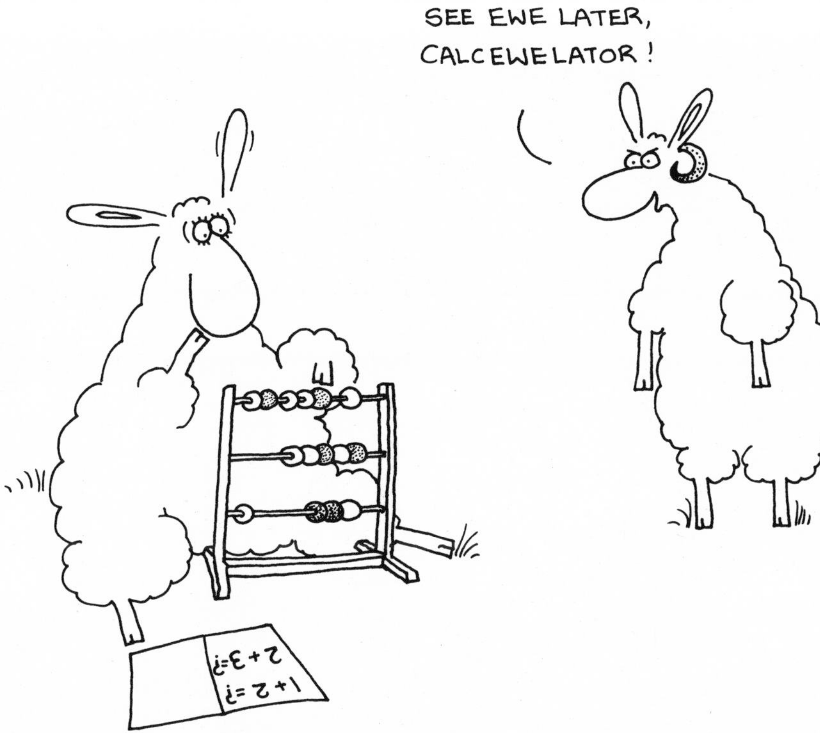
SEE EWE LATER,
CALCEWELATOR !

ABAACUS

HANDLEBAA MOUSTACHE

EWE-LIPS FROM AMSTERDAM

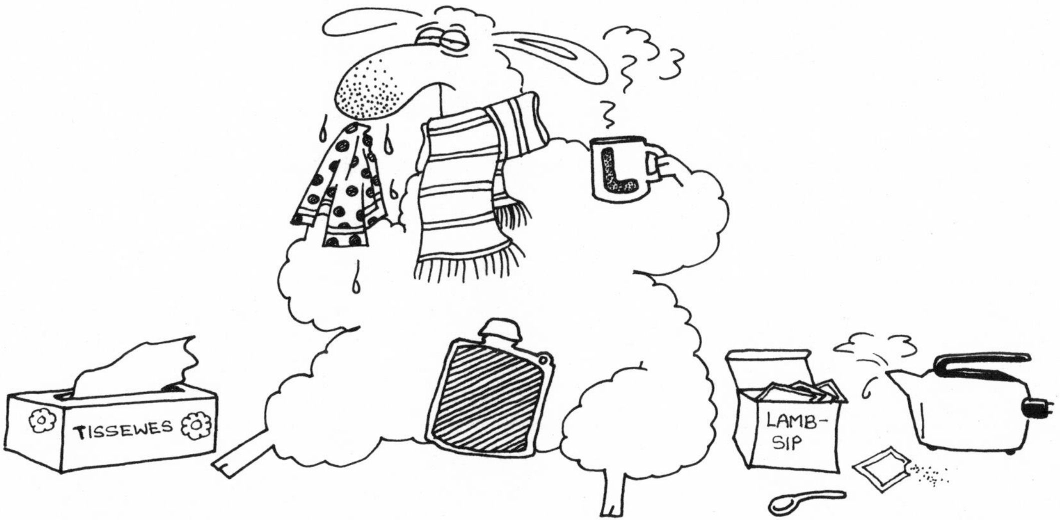
TISSEWES
LAMB-
SIP

LAMB SIP

CURSE EWE, RED BAARON !

THE SHEEP OF THINGS TO COME

ALI BAA BAA

SHEEPS THAT PASS IN THE NIGHT

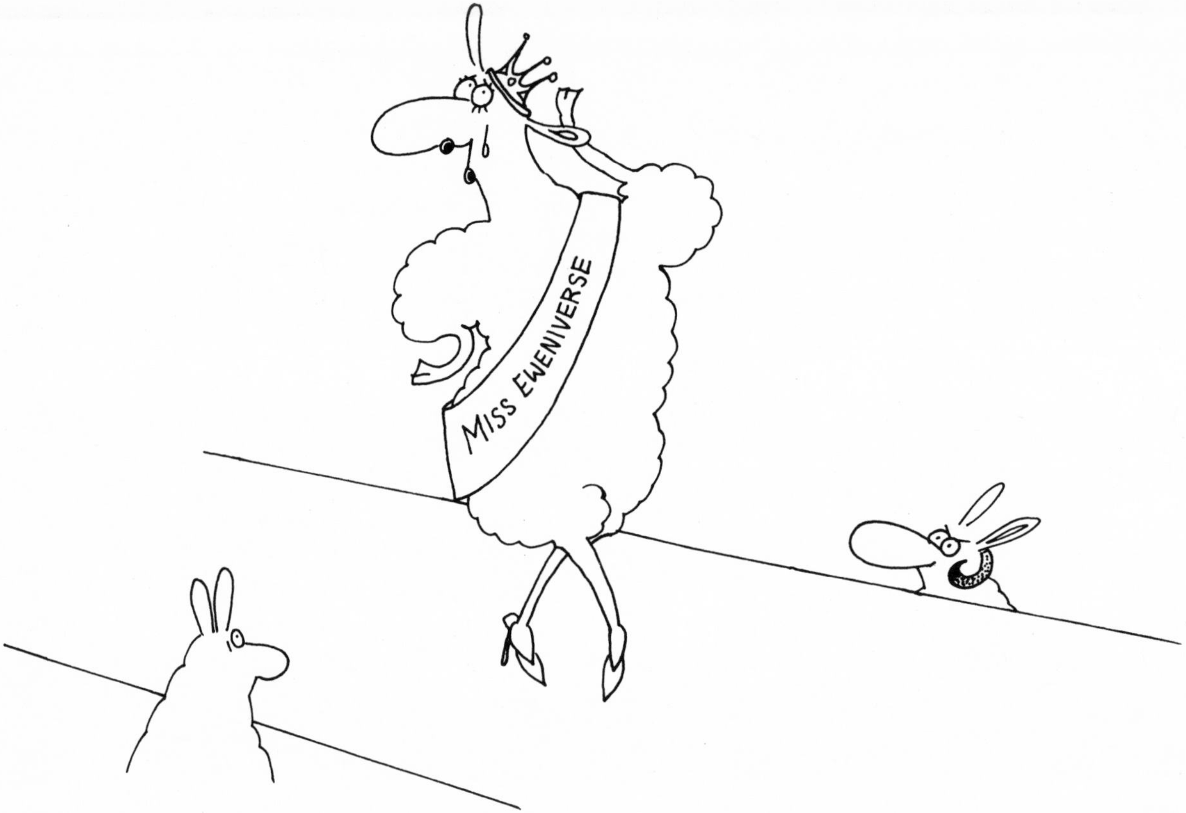
MISS EWENIVERSE

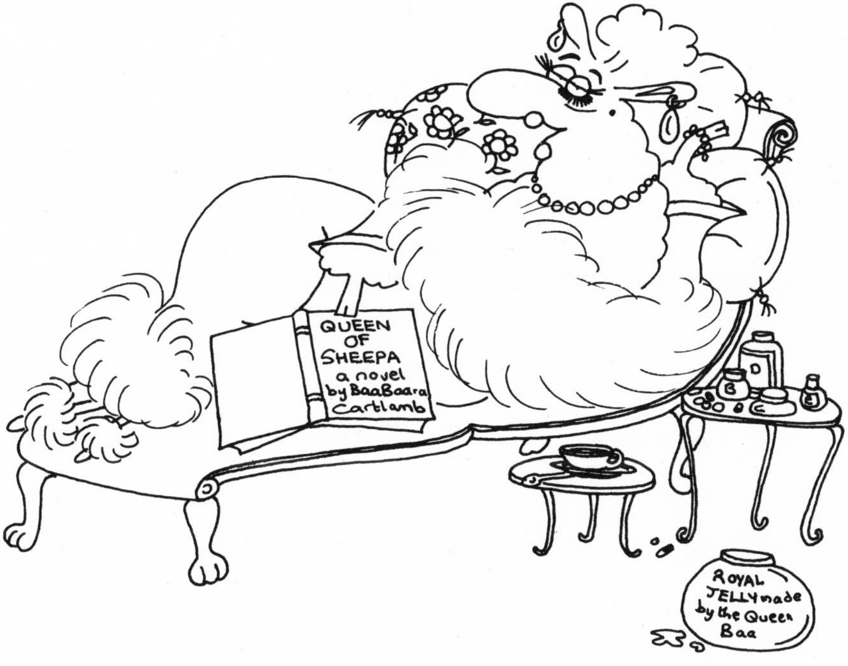
QUEEN OF SHEEPA a novel by BaaBaara Cartlamb
ROYAL JELLY made by the Queen Baa

BAABAARA CARTLAMB

RUM BAA BAA

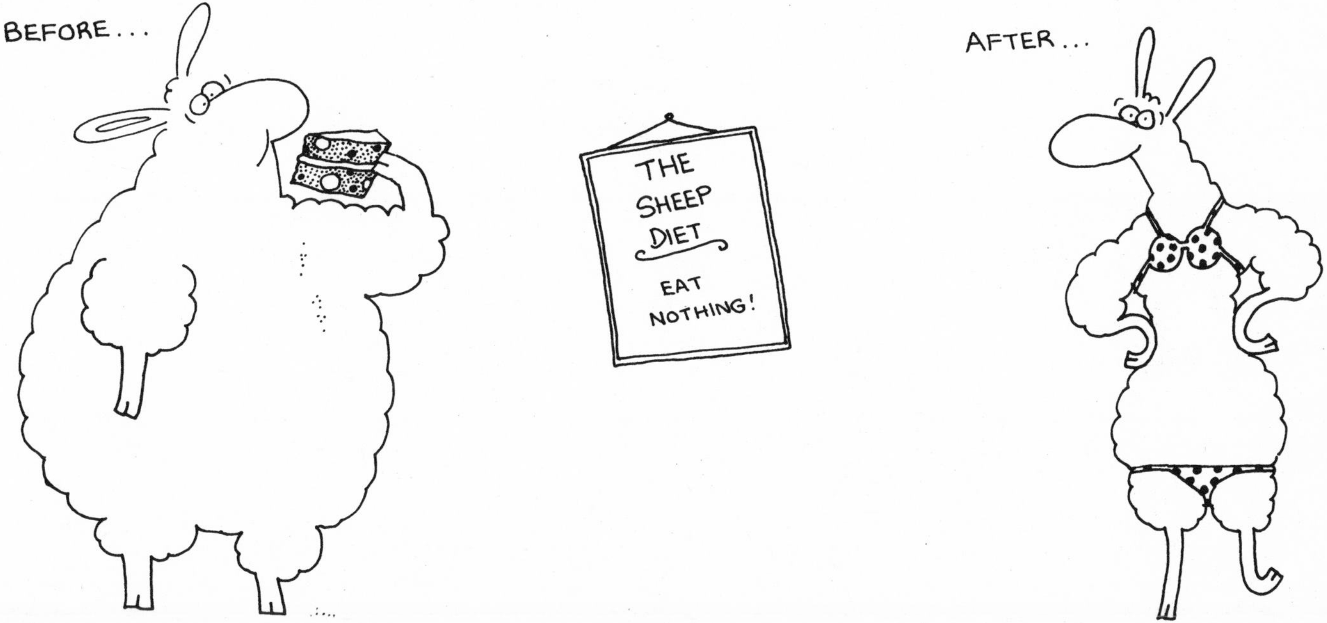
BEFORE...
THE SHEEP DIET
EAT NOTHING!
AFTER...

SHEEP SHED

Cock and Bull Stories

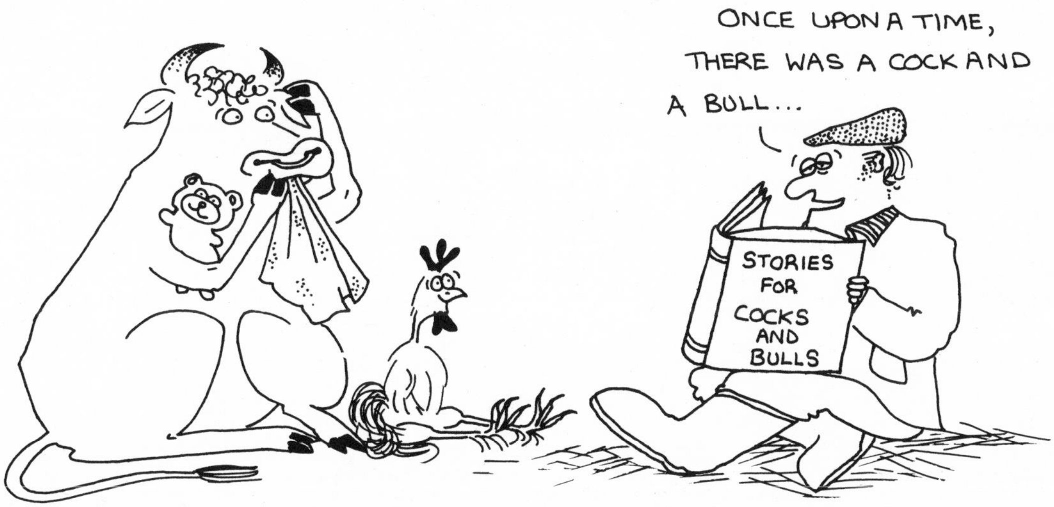
ONCE UPON A TIME, THERE WAS A COCK AND A BULL...
STORIES FOR COCKS AND BULLS

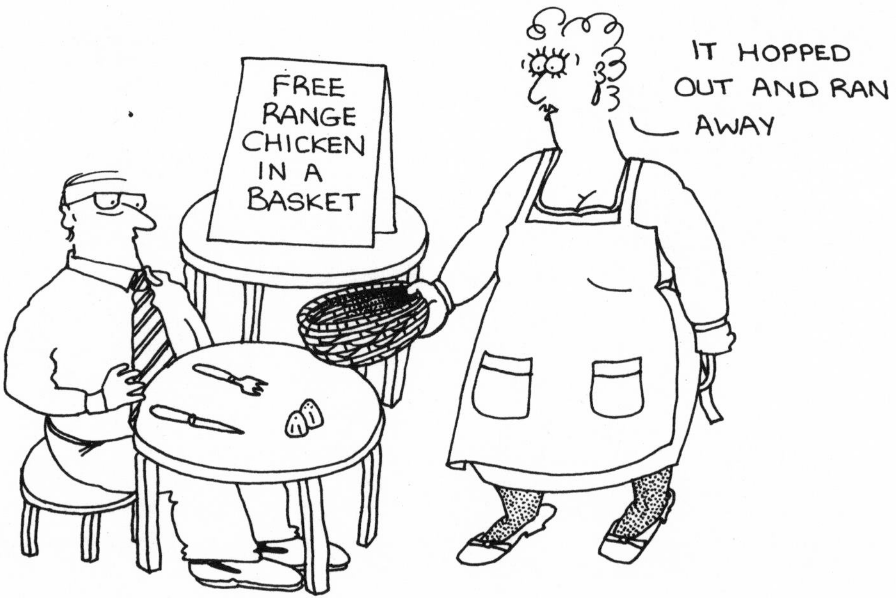
FREE
RANGE
CHICKEN
IN A
BASKET
IT HOPPED
OUT AND RAN
AWAY

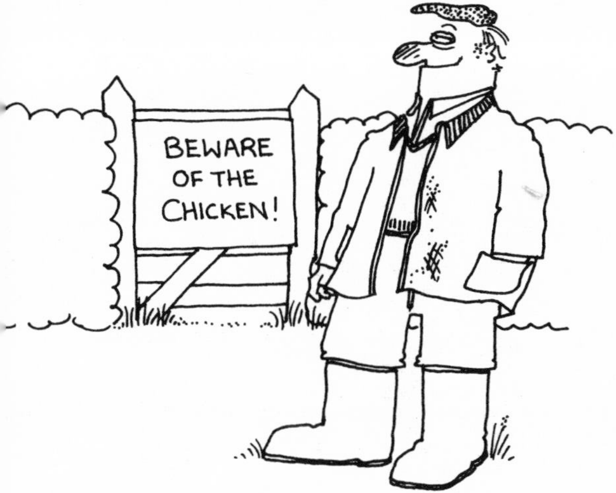
BEWARE
OF THE
CHICKEN!

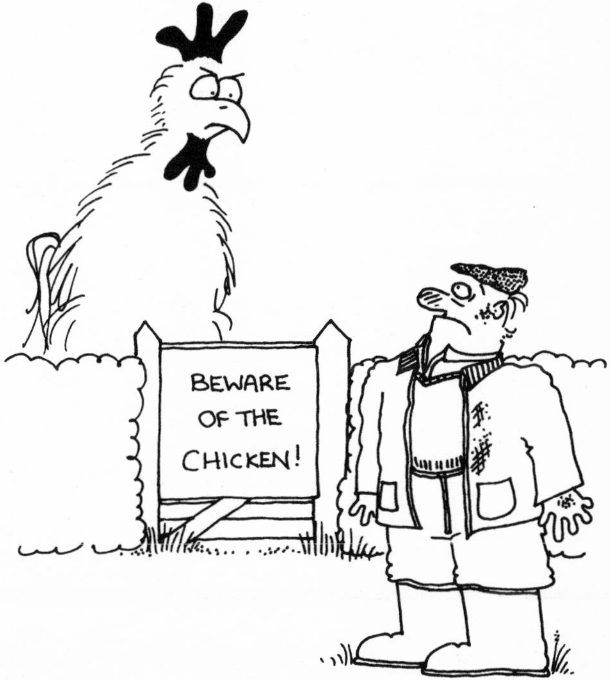
BEWARE
OF THE
CHICKEN!

SCOTCH EGG   COCK — A — LEEKIE

52

POACHED EGG

53

COCK - AU - VIN

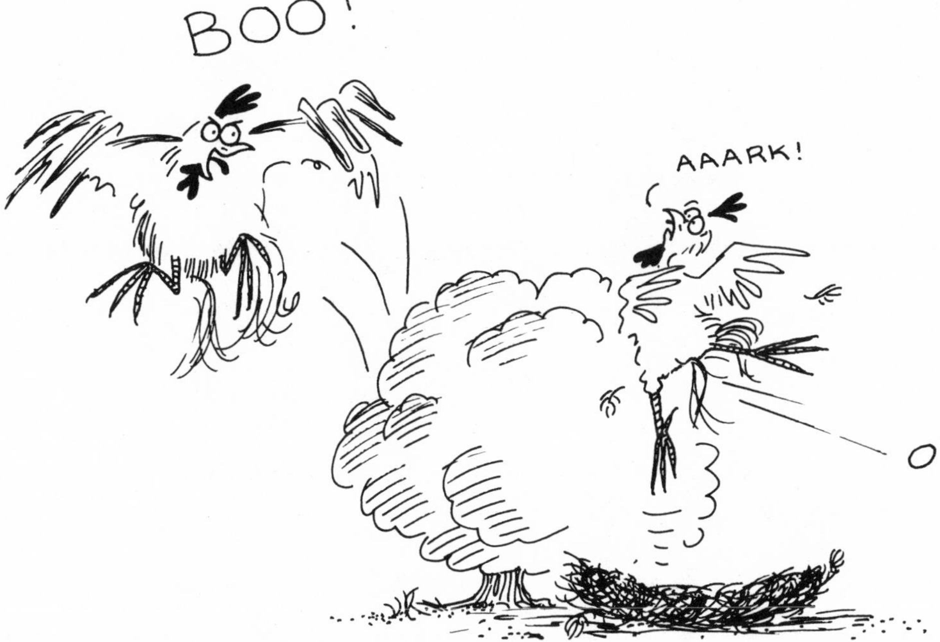
BOO!
AAARK!

FOWL PEST

BALL COCK

COCK - A - HOOP

MAD TURKEY DISEASE! MAD TURKEY DISEASE! UNCLEAN! UNCLEAN!
IT'LL NEVER WORK

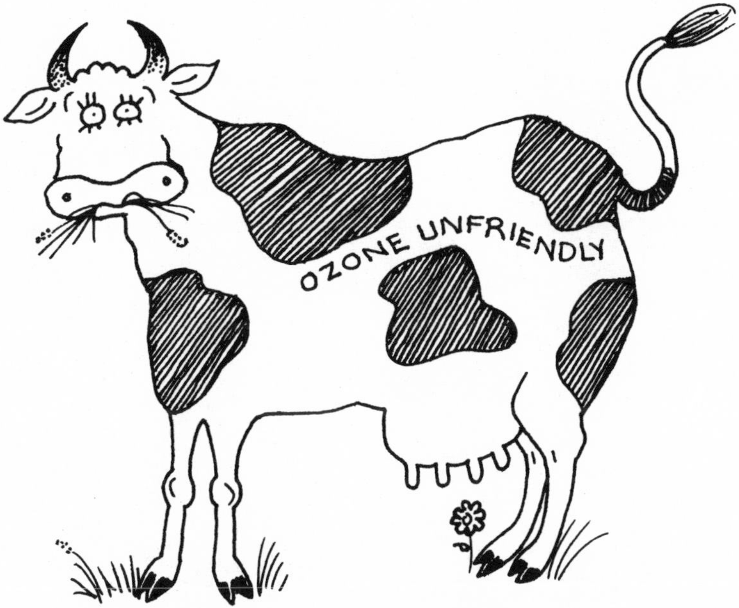
OZONE UNFRIENDLY

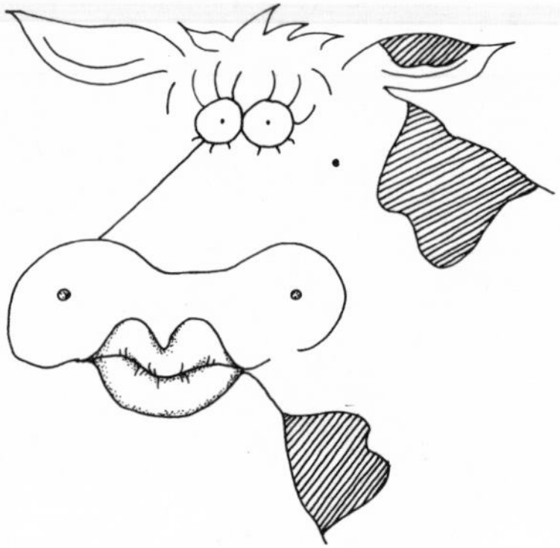

COWSLIPS

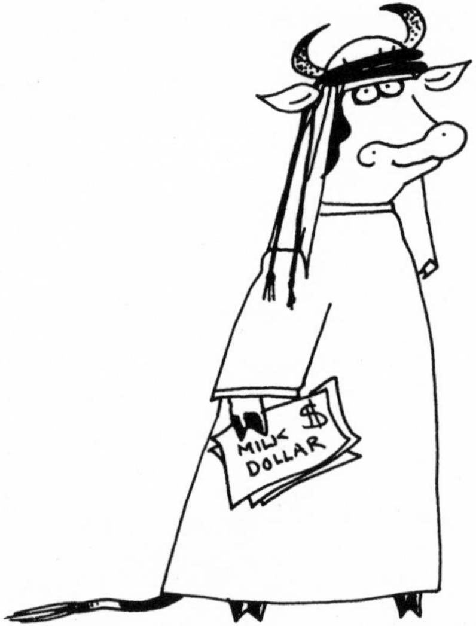
$
MILK DOLLAR

MILK SHEIKH

MAKING BUTTER

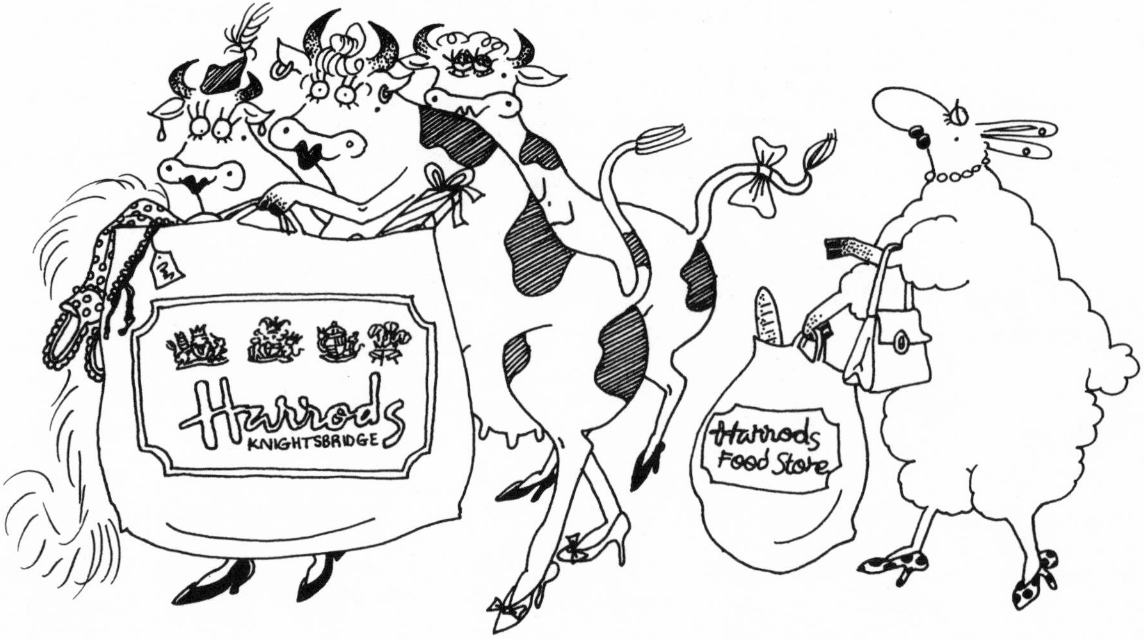
Harrods
KNIGHTSBRIDGE
Harrods
Food Store

STORE CATTLE AND STORE SHEEP

I'M FRIESIAN!
I'M NOT TOO WARM MYSELF!

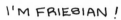

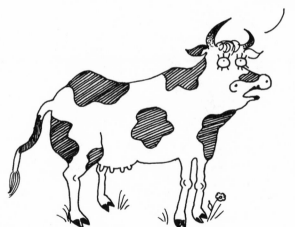

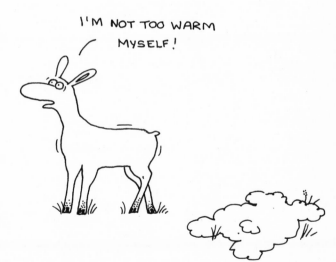

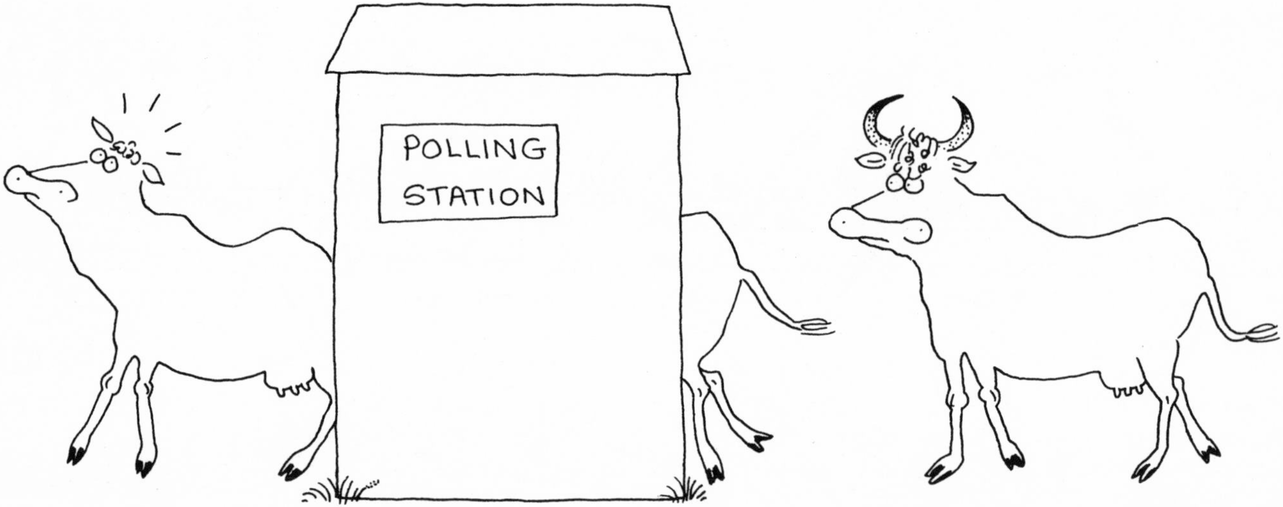
POLLING
STATION

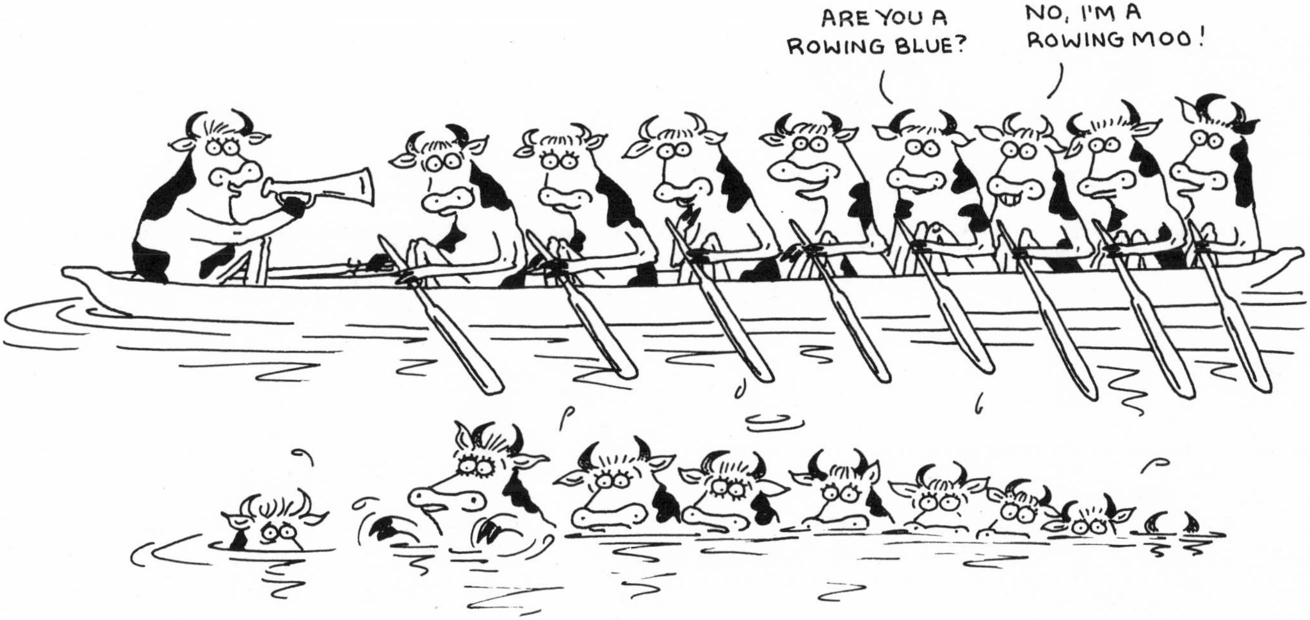
ARE YOU A ROWING BLUE?
NO, I'M A ROWING MOO!

OXFORD AND COWBRIDGE

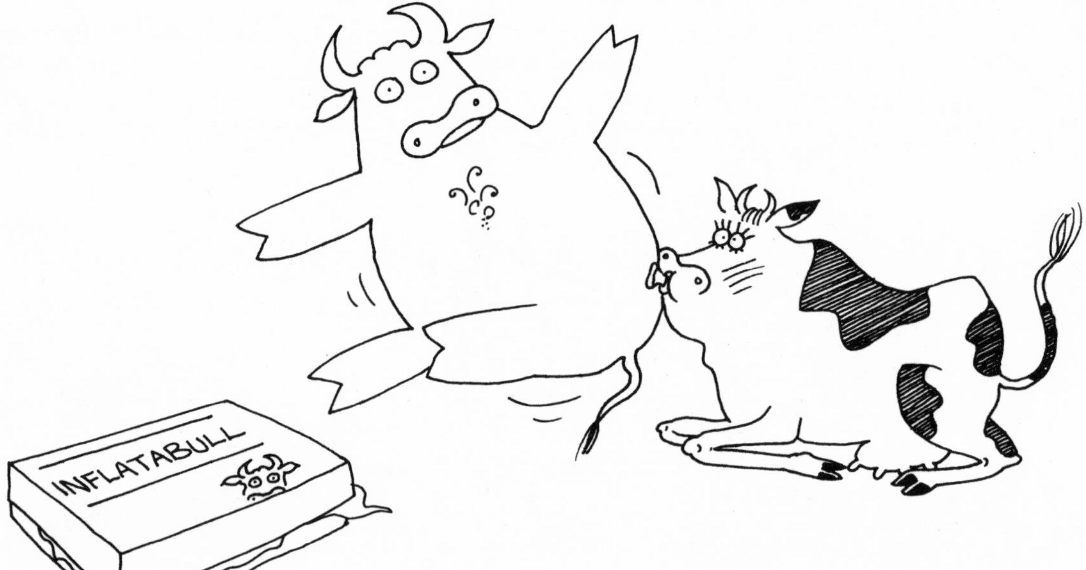
INFLATABULL

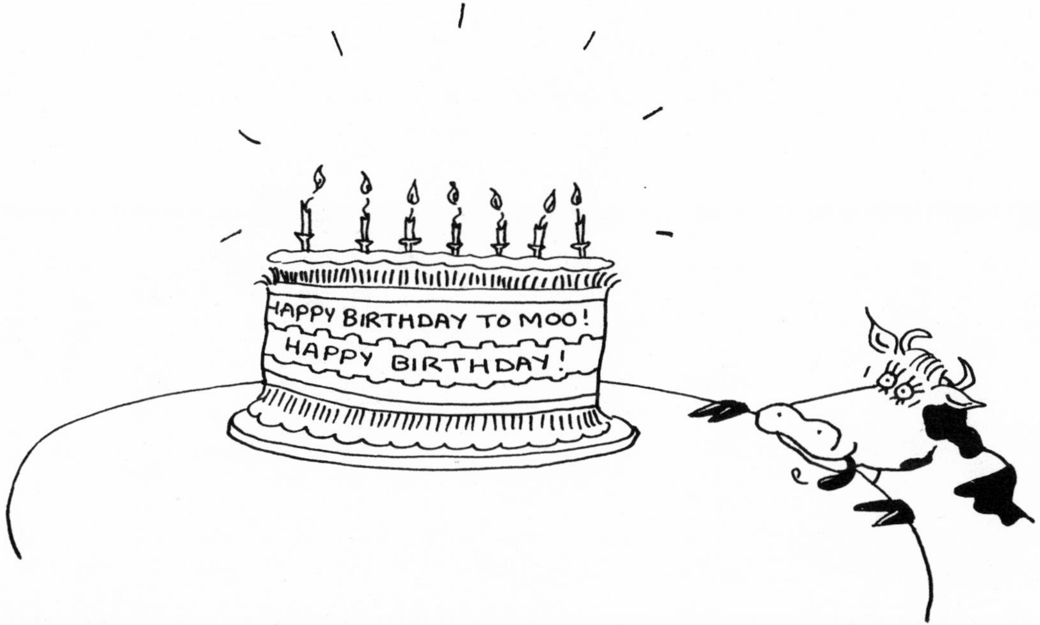
HAPPY BIRTHDAY TO MOO!
HAPPY BIRTHDAY!

CATTLE CAKE

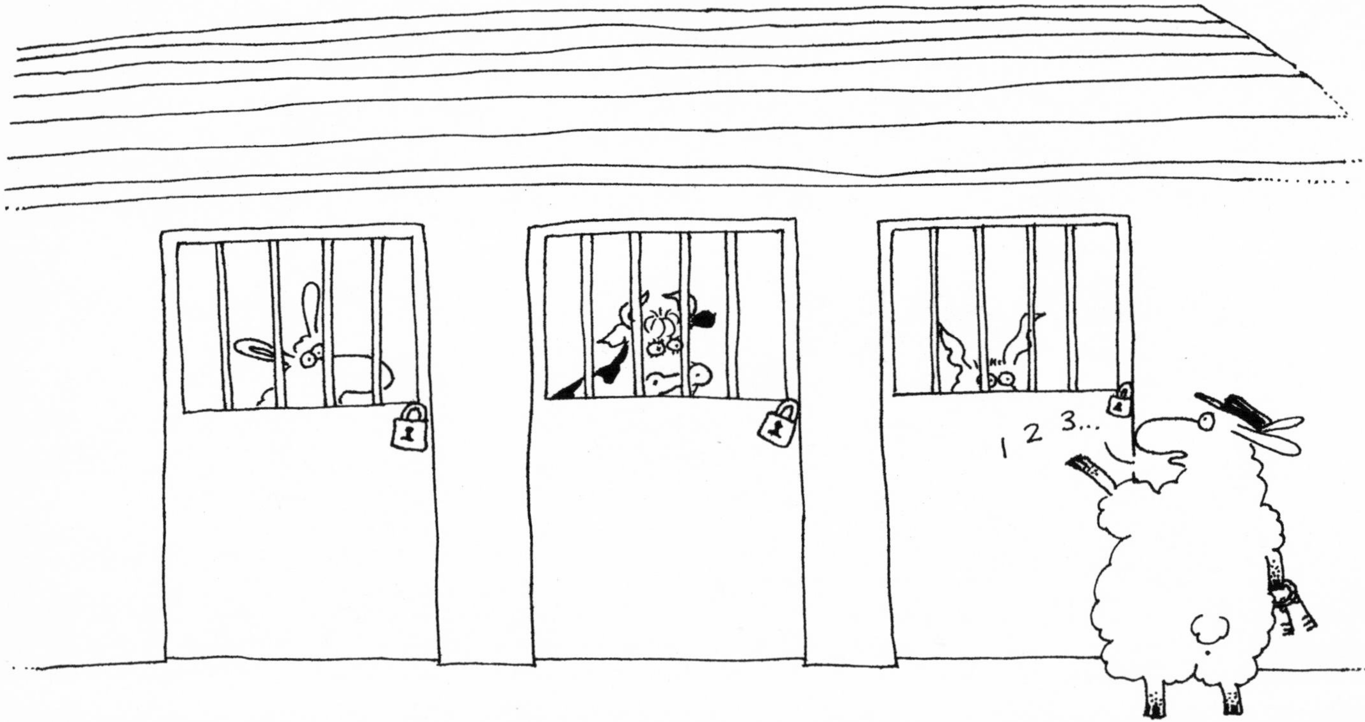
1 2 3...

CELL COUNT

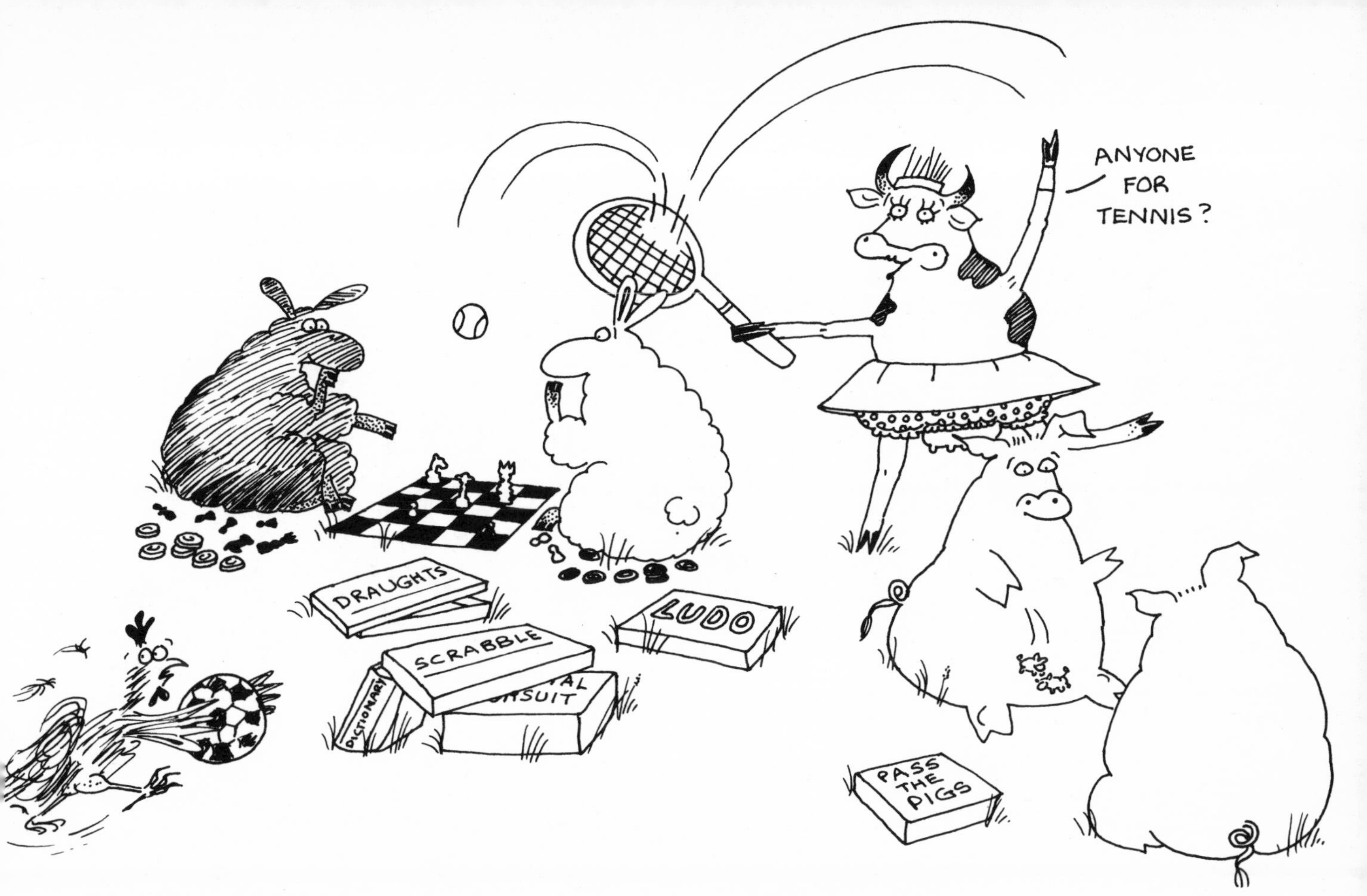
ANYONE FOR TENNIS?
DRAUGHTS
SCRABBLE
LUDO
DICTIONARY
PASS THE PIGS

START OF THE GAME SEASON

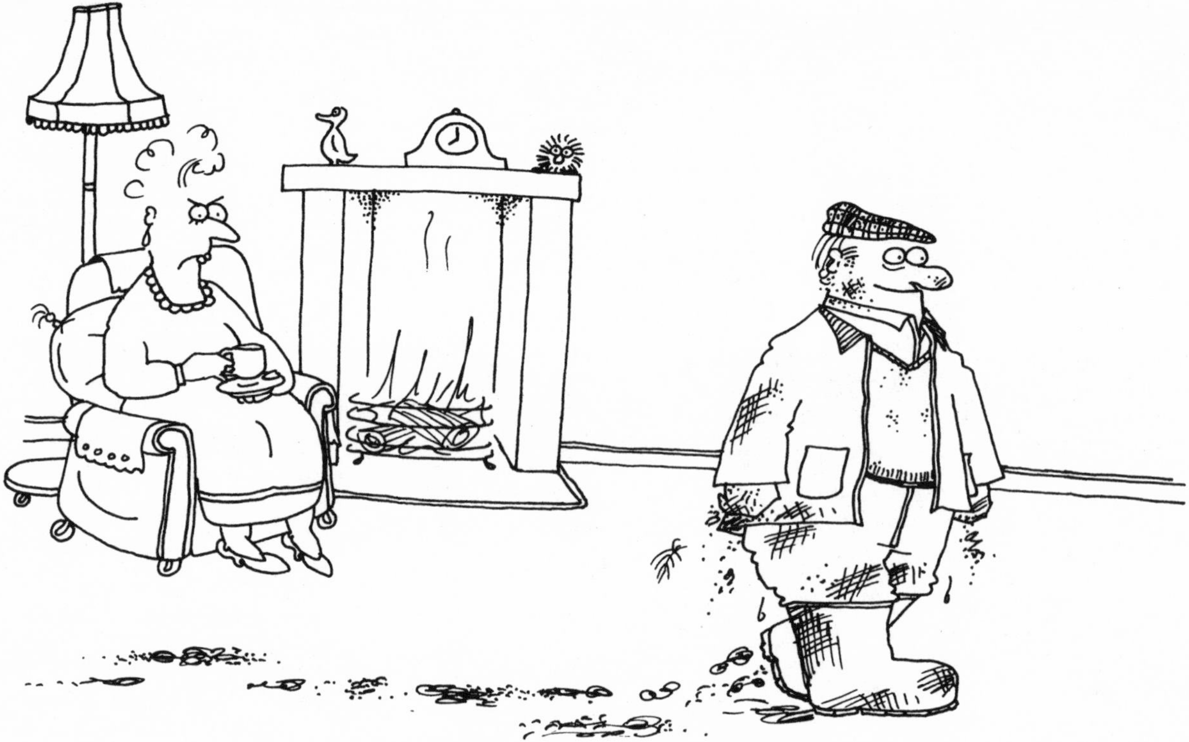

MUCK – SPREADING

Hogwash

I'VE NEVER BEEN MUCH GOOD AT SMALL PORK!
MOTHER AND PIGLET
YOUR LITTER AND YOU

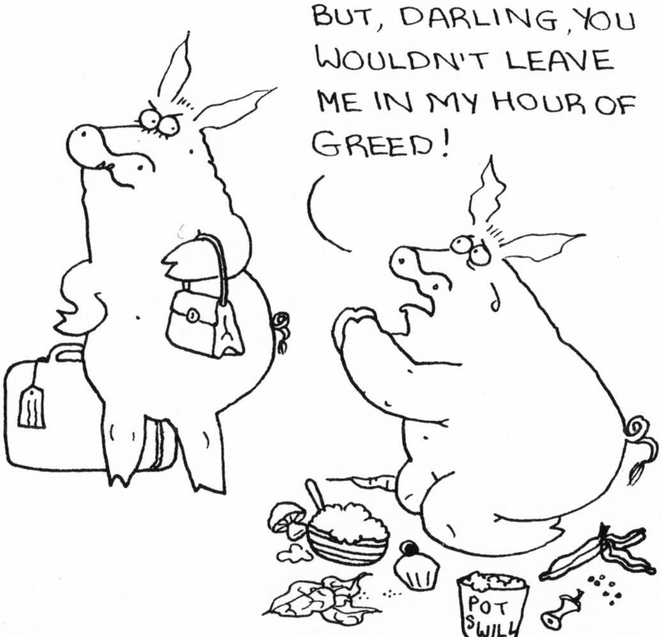
BUT, DARLING, YOU WOULDN'T LEAVE ME IN MY HOUR OF GREED!
POT SWILL

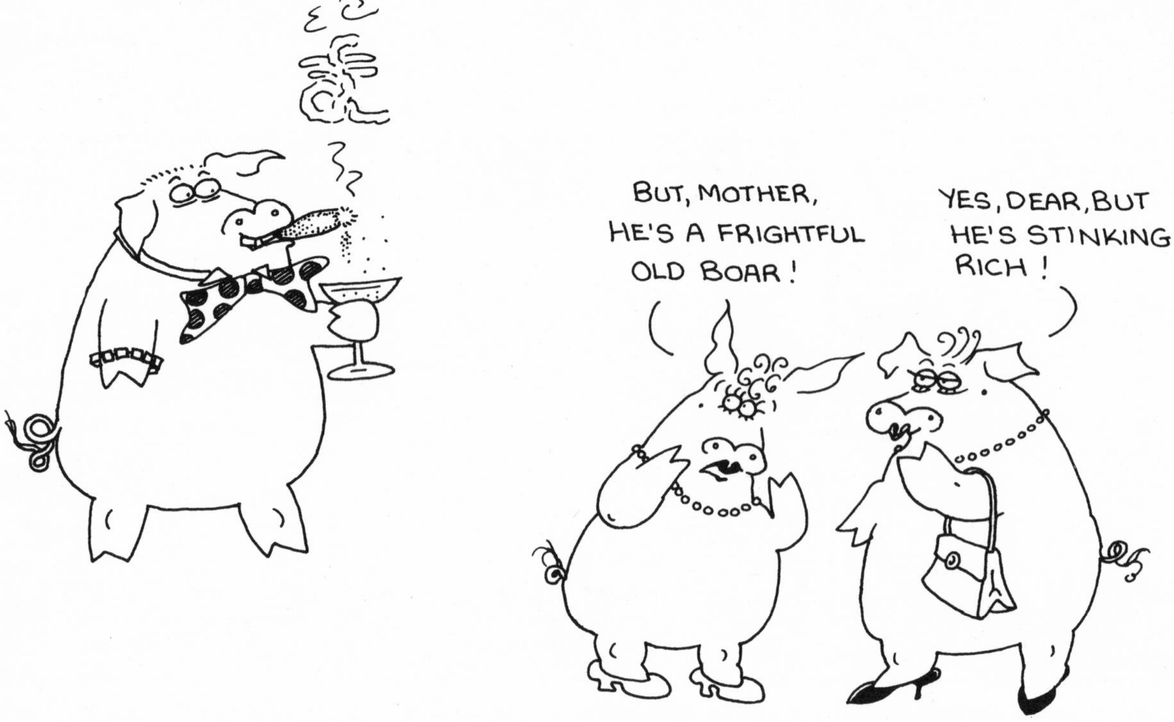
BUT, MOTHER, HE'S A FRIGHTFUL OLD BOAR!
YES, DEAR, BUT HE'S STINKING RICH!

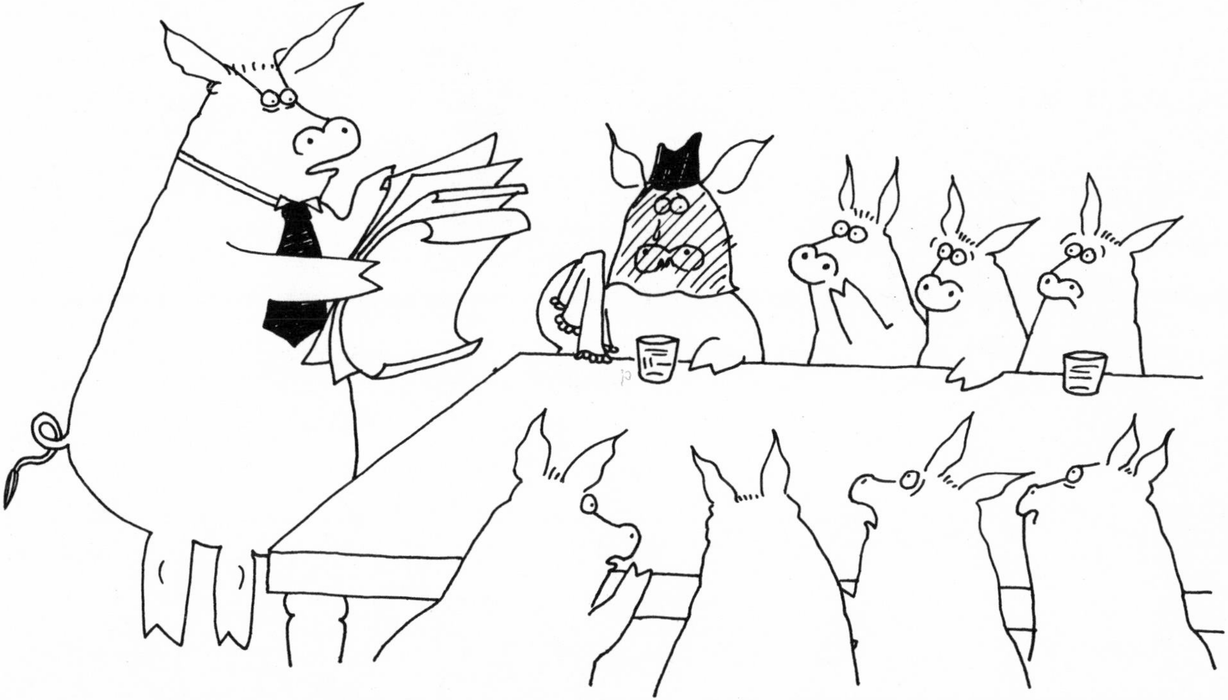

PIG'S WILL

IMPORKUNIOUS

PORK OF THE TOWN

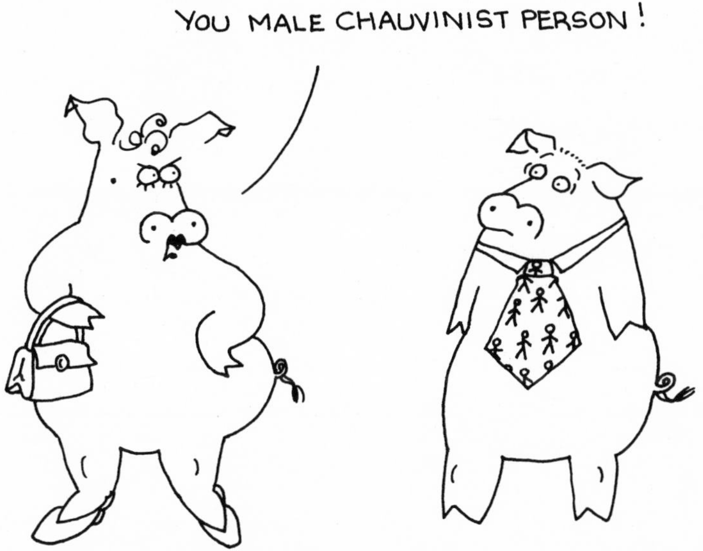
YOU MALE CHAUVINIST PERSON!

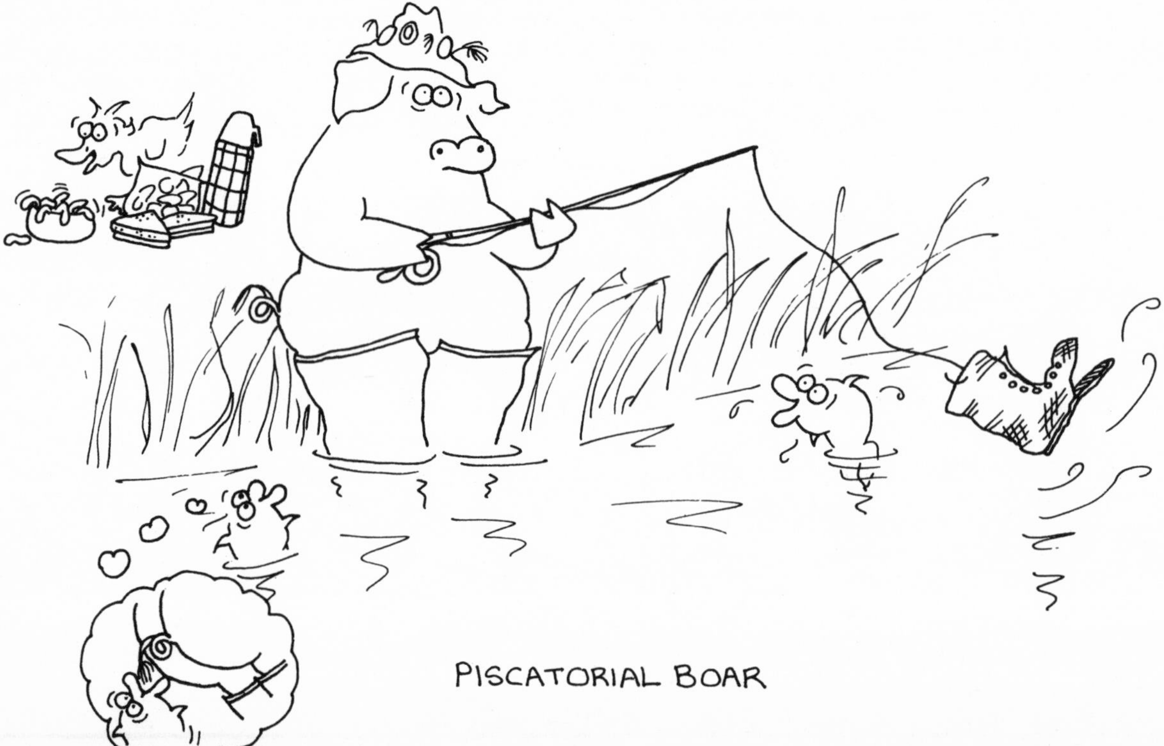

PISCATORIAL BOAR

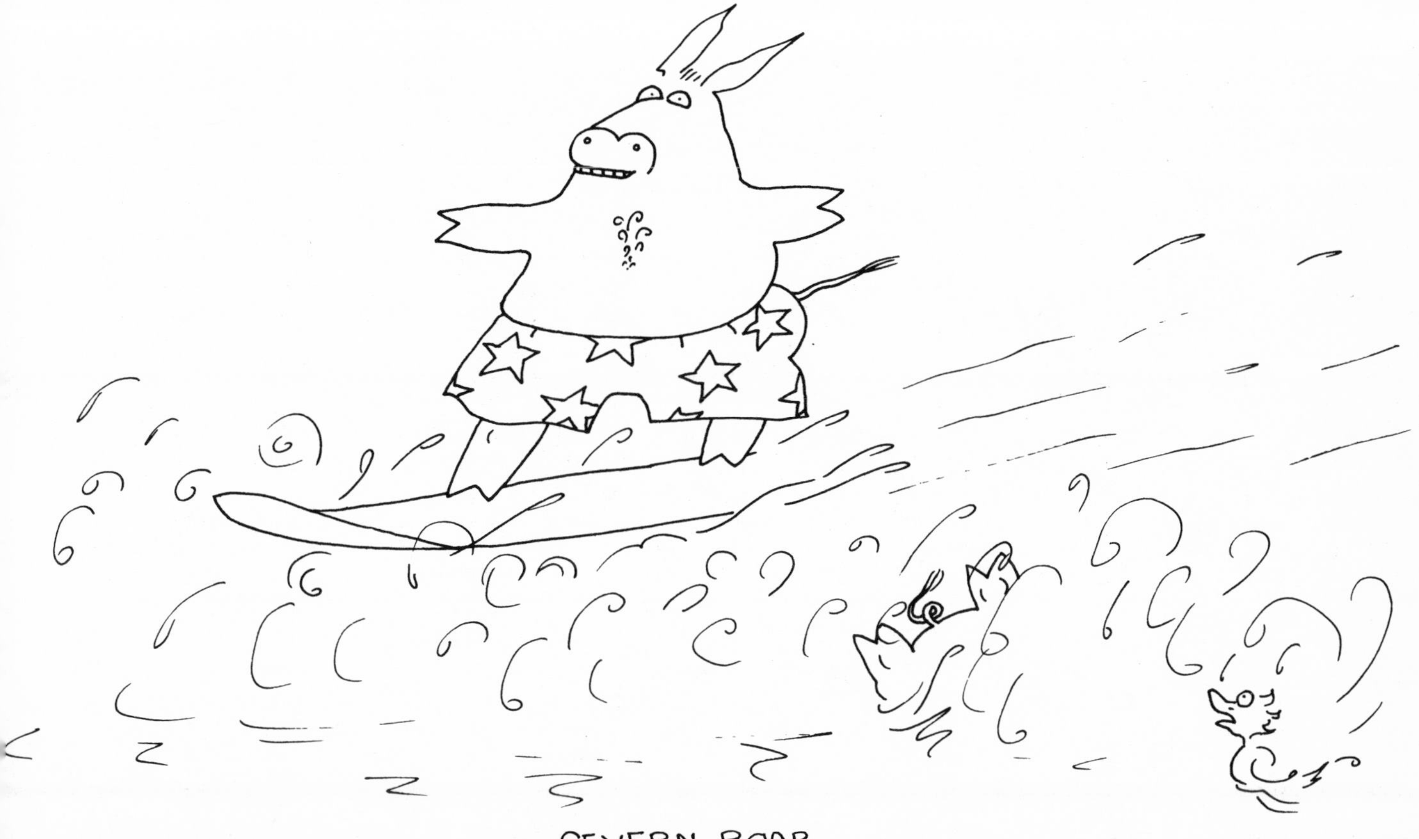

SEVERN BOAR

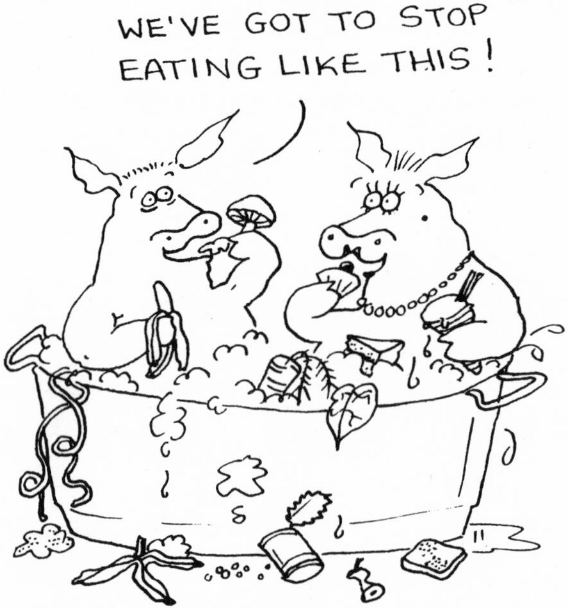
WE'VE GOT TO STOP EATING LIKE THIS!

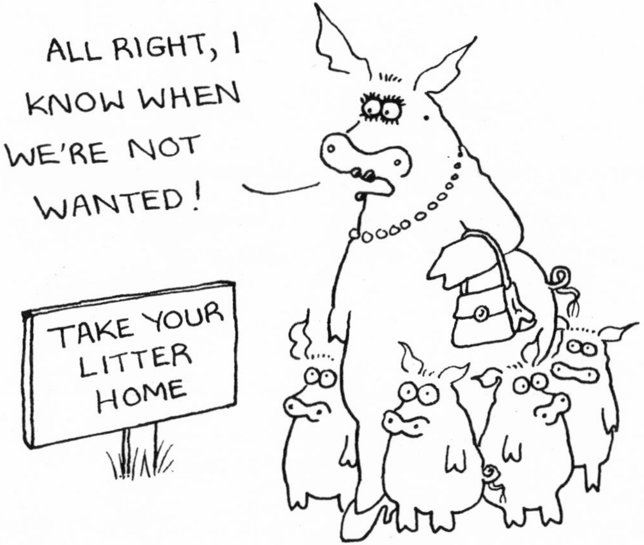
ALL RIGHT, I KNOW WHEN WE'RE NOT WANTED!
TAKE YOUR LITTER HOME

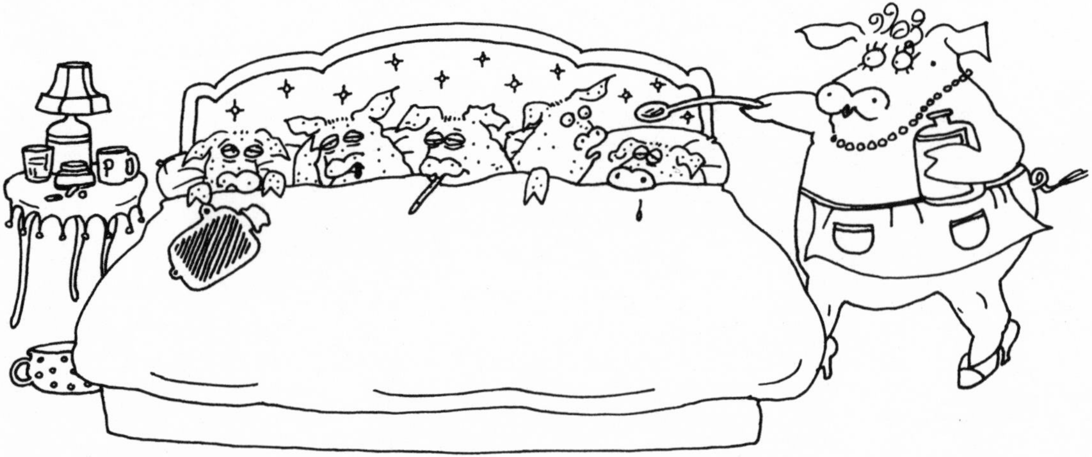
P U

LITTERBUG

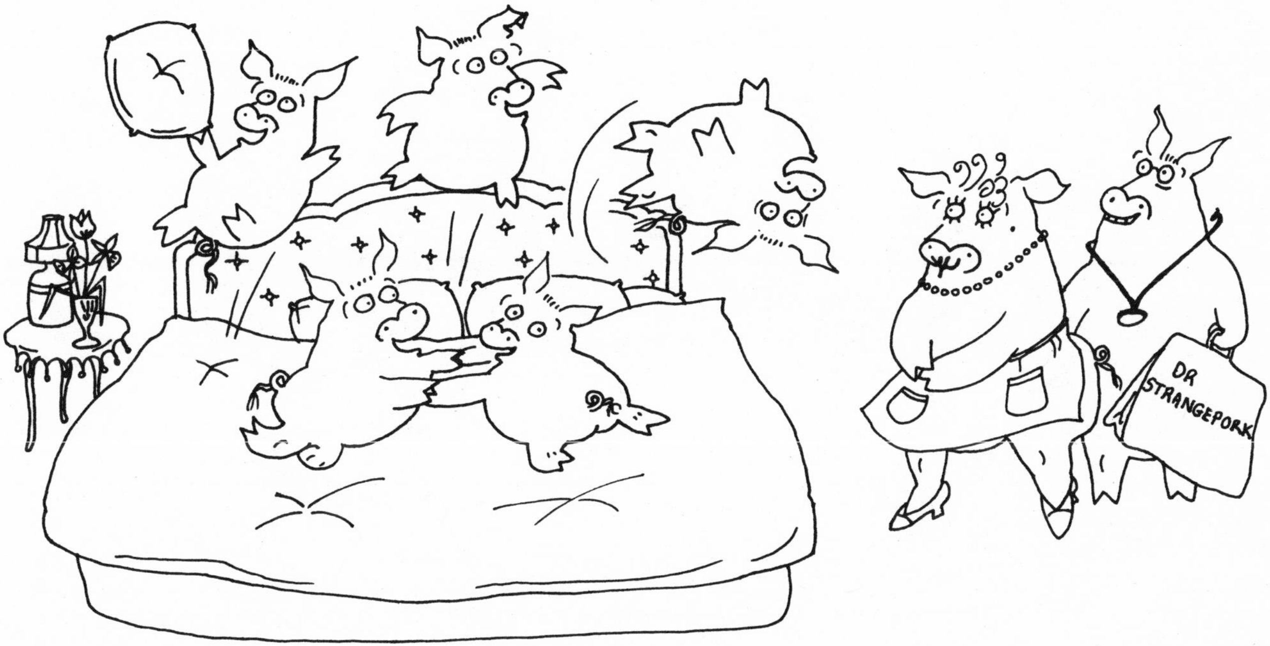
DR STRANGEPORK

CURED BACON

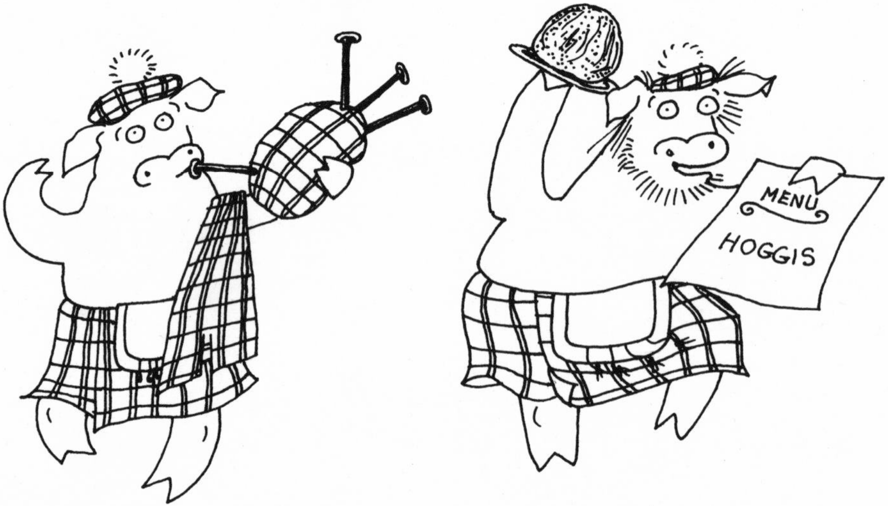
MENU
HOGGIS

HOGMANAY

Z
Z
Z
W
W

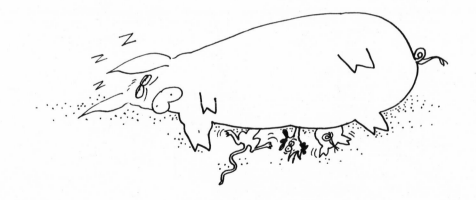

IDLE PORK COSTS LIVES

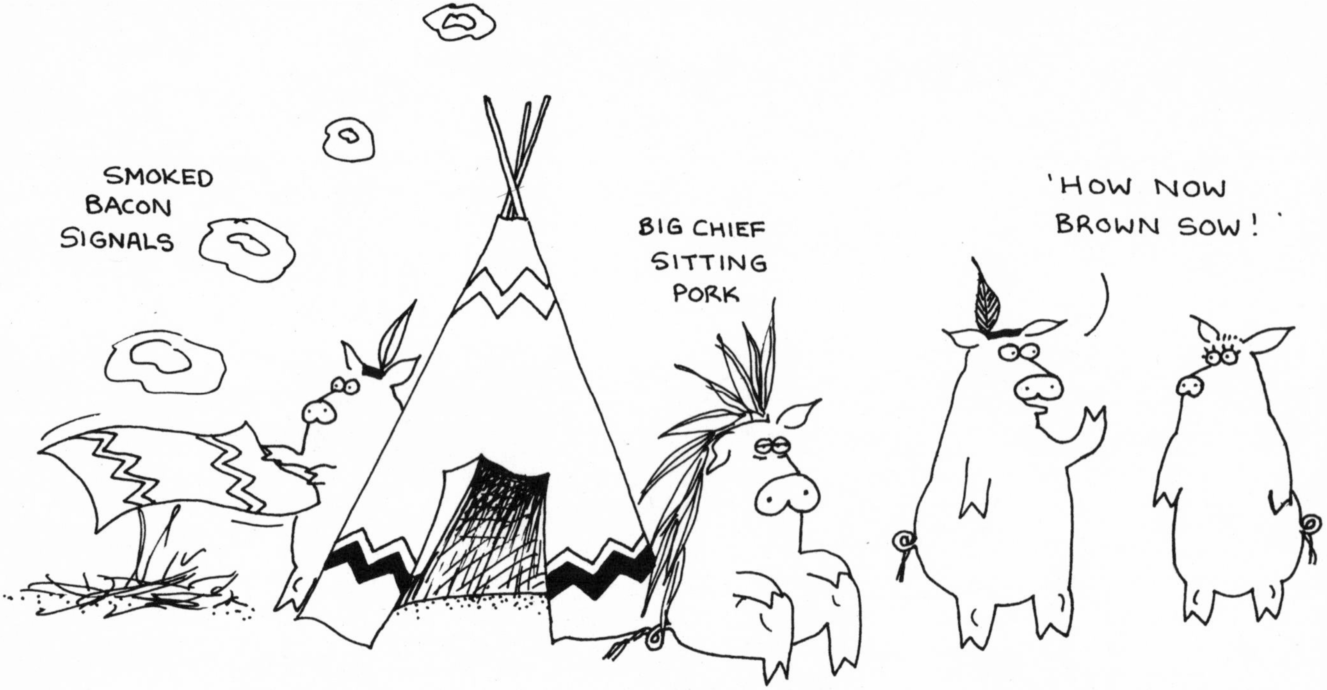
SMOKED
BACON
SIGNALS
BIG CHIEF
SITTING
PORK
'HOW NOW
BROWN SOW!'

PIG-WAM

A LOW THRESHOLD OF BOARDOM

PIG'S TIE   PORK SCRATCHINGS

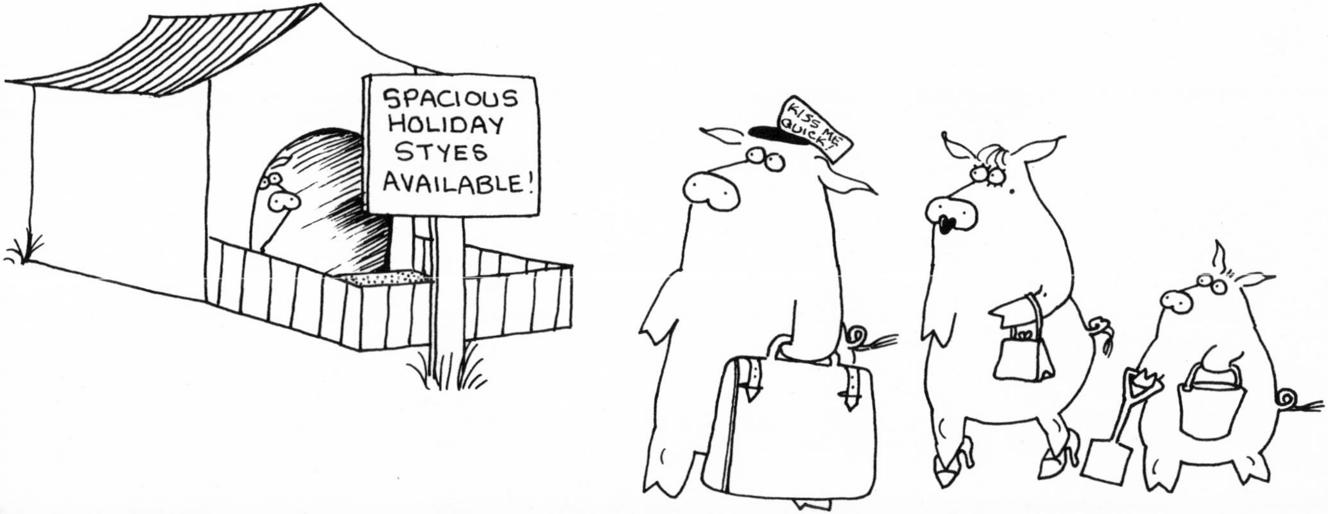
SPACIOUS
HOLIDAY
STYES
AVAILABLE!
KISS ME
QUICK!

PIG LET

KATHRYN LAMB

Kathryn Lamb began working as a cartoonist for *Private Eye* magazine while she was an English literature student at Oxford. She has illustrated children's books and exhibits her sheep cartoons at London's Cartoon Gallery.

Kathryn grew up in the Middle East where her father was British Ambassador to Kuwait. She has three children and three stepchildren and lives in Dorset.

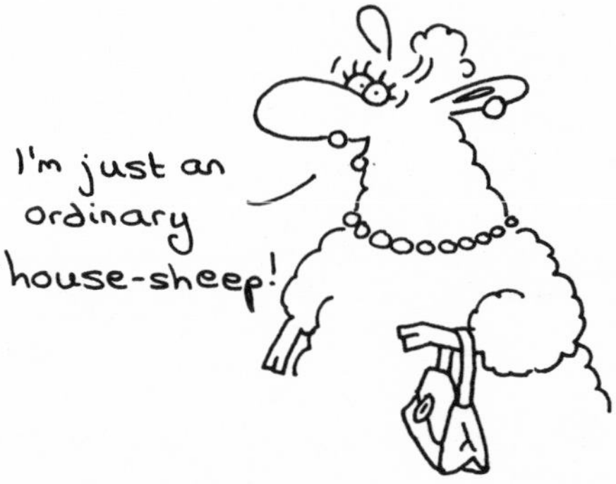
I'm just an ordinary house-sheep!

FARMING PRESS BOOKS

Other humorous authors published by Farming Press include:

Henry Brewis
Based on his farming experience in Northumberland, cartoons, verses and stories featuring bank managers, reps, sheepdogs, tourists and other plagues on the hill farmer's life
- *Clarts and Calamities*
- *Chewing the Cud*
- *Don't Laugh Till He's out of Sight*
- *Funnywayt'mekalivin'*
- *The Magic Peasant*

Veronica Frater
How to cope with a farmhouse accommodation and seven young children, and stay more or less sane
- *They All Ran After the Farmer's Wife*

John Terry
From school dump to model school farm and then on to success in the show ring
- *Calves in the Classroom*
- *Ducks in Detention*
- *Pigs in the Playground*

Emil van Beest
Preposterous cartoons about farm animals
- *Fowl Play*
- *Pigmania*
- *Udderwise*

James Robertson
Classic accounts of an innocent stumbling into agriculture
- *Any Fool Can Be A Dairy Farmer*
- *Any Fool Can Be A Pig Farmer*

For more information or a free illustrated book list please contact:

Farming Press Books, 4 Friars Courtyard
30-32 Princes Street, Ipswich IP1 1RJ, United Kingdom
Telephone (0473) 241122